AN AMERICAN GEOLOGIST'S ADVENTURES IN AFRICA

A ROCK AND A HARD PLACE

GEORGE ZELT, PhD

GREENLEAF
BOOK GROUP PRESS

The names and identifying characteristics of persons referenced in this book, as well as identifying events and places, have been changed to protect the privacy of the individuals and their families.

Published by Greenleaf Book Group Press
Austin, Texas
www.gbgpress.com

Distributed by Greenleaf Book Group

For ordering information or special discounts for bulk purchases, please contact Greenleaf Book Group at PO Box 91869, Austin, TX 78709, 512.891.6100.

Design and composition by Greenleaf Book Group
Cover design by Greenleaf Book Group
All photos graciously provided by George and Zhanar Zelt
Cover images: Marble rocks on Marble Beach in Namaqua National Park, South Africa, Victoria Martin; Brushed metal texture used under license from Shutterstock.com ©iStockphoto.com/xxmmxx; ©iStockphoto.com/Freder
Image Credits Interior: Bushmen (san) rock painting of antelopes, ©iStockphoto.com/EcoPic and Prehistoric Bushman engravings at Twyfelfontein in Namibia, ©iStockphoto.com/PytyCzech

Publisher's Cataloging-in-Publication data is available.

Print ISBN: 978-1-62634-571-3

eBook ISBN: 978-1-62634-572-0

Part of the Tree Neutral® program, which offsets the number of trees consumed in the production and printing of this book by taking proactive steps, such as planting trees in direct proportion to the number of trees used: www.treeneutral.com

Printed in the United States of America on acid-free paper

19 20 21 22 23 24 10 9 8 7 6 5 4 3 2 1

First Edition

To my parents, Marie D. and George E. Zelt,
and brother Norman F. Drum. With me from the
beginning and gone too soon, I remember them always
for their understanding, support, and love.

Still round the corner there may wait
A new road or a secret gate . . .

—"A Walking Song" from *The Lord of the Rings* by J.R.R. Tolkien

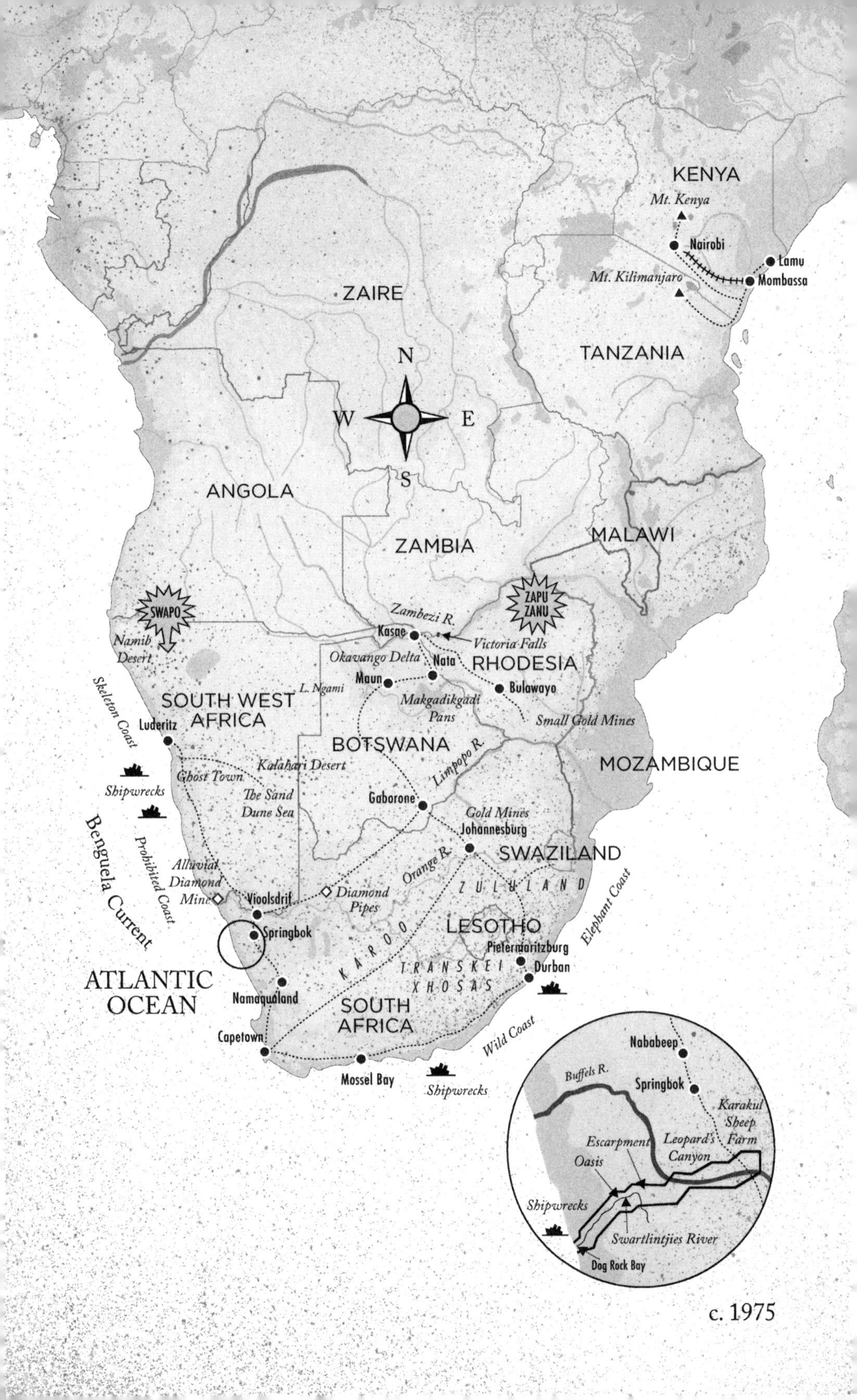

KENYA
Mt. Kenya
Nairobi
Lamu
Mombassa
Mt. Kilimanjaro
ZAIRE
TANZANIA
N
W
E
S
ANGOLA
ZAMBIA
MALAWI
SWAPO
ZAPU
ZANU
Zambezi R.
Kasae
Victoria Falls
Namib Desert
Okavango Delta
Nata
RHODESIA
Maun
Bulawayo
Skeleton Coast
L. Ngami
SOUTH WEST AFRICA
Makgadikgadi Pans
Small Gold Mines
Luderitz
BOTSWANA
MOZAMBIQUE
Kalahari Desert
Limpopo R.
Ghost Town
Shipwrecks
The Sand Dune Sea
Gaborone
Gold Mines
Johannesburg
Benguela Current
Probibited Coast
SWAZILAND
Alluvial Diamond Mine
Orange R.
Elephant Coast
Vioolsdrif
Diamond Pipes
ZULULAND
Springbok
KAROO
LESOTHO
Pietermaritzburg
TRANSKEI
Durban
ATLANTIC OCEAN
XHOSAS
Namaqualand
SOUTH AFRICA
Capetown
Wild Coast
Mossel Bay
Shipwrecks
Nababeep
Buffels R.
Springbok
Karakul Sheep Farm
Escarpment
Leopard's Canyon
Oasis
Shipwrecks
Swartlintjies River
Dog Rock Bay
c. 1975

Contents

Chapter 1

1977: Prohibited Area, South West Africa

"How long, Patrick?" I asked, squinting through the back window of our Land Rover.

"Five, maybe six minutes." His keen eyes darted from side to side as we raced over the immense expanse of sand. "Then, bejesus, it's on us . . . a frantic beast with whirling sand claws . . . It will blot out the sun as it beats the hell out of us."

"That's kind of a poetic description," I replied, surprised and turning toward him. He wore his white long-sleeved shirt hanging out of his jeans, as did I. White helped reflect the omnipotent sun's assault on the desolate land below. Steamy heat waves rose, quivering. We were in our adventurous twenties, agile and fit, but now worried.

My throat tightened, and the heat increased as we slid the Land Rover's windows closed. There was no room for error when the desert bowled a sandstorm. It could kill. At six foot three, I could look out the top of the boxy windshield while Patrick stared intensely out the middle and drove.

Where should we go?

It was all so very empty.

Wide-brimmed hats lay on the seat next to us. Under them were rather featureless maps and a Brunton compass, the type we geologists carried. For a map to be of use, we needed topography. But there wasn't much of that out here.

Tapping sand fingertips began to play on our metal roof. Grains seeped inside through passages unknown, peppering Patrick's jet-black hair and scruffy beard, then blending with the sun-bleached strands hanging from my head. Fine particles began to work their way into our ankle boots and pants. We had been traveling some days through the Namib Desert; our luck had run out.

Within moments the darkening gray cloud behind us increased in size, rolling in a wave of fury as it sucked air and sand into its lungs. The bright desert was metamorphosing, extracting payment for our intrusion. The ungainly Land Rover plowed forward.

We were in the vast Sperrgebiet, or prohibited region, of South West Africa (which in 1990 would become Namibia), geology students illegally hunting for diamonds.

The area bordered the Atlantic Ocean, stretching north from the Orange River border of South Africa for about two hundred miles and extending some sixty miles inland. It was also known as Diamond Area 1 by then-owner Consolidated Diamond Mines of South West Africa Ltd., which patrolled it using men with guard dogs.

As our Land Rover charged up a gently rolling rise, Patrick pointed through the haze in front of us to a long, isolated ridge of massive rock in the near distance.

"Precambrian," Patrick yelled over the increasing wind. "A billion years old." The ridge jutted upward like the desert's spine. Maybe we could find a hole between rocks to crawl into.

We gained speed. Lances of sand struck our roof furiously as the Rover lunged and finally reached the apron of boulders strewn along the base of the rock wall.

"I'll drive along by the boulders!" Patrick yelled. "Look for a crack in the rocks, a hollow, some protection!"

The tuneless drumming increased. We inhaled wisps of the storm's desiccated, heated breath. No vehicle could keep it out. We tied our well-used bandanas around our faces and looked like the thieves we were. Moments later we could see only twenty meters ahead as the sun's rays struggled through a sand army with trillions of airborne soldiers.

A hazy black hole just big enough to squeeze through appeared in the gray granite-gneiss ridge wall we were passing. Patrick braked and maneuvered into the boulders in front of it. His clenched hands twisted the steering wheel as if to strangle it. I slid back my window to watch for angled rocks that could slit our tires.

"This okay?" he yelled anxiously, unable to see much. The motor gasped for want of air while the wind whistled eerily as it tore along the ridge and between its fallen boulders.

"Hell," I replied, bringing my head back inside, covered with sand. "No choice! Shut it off."

Securing our hats with chinstraps, we clambered out, dragging a canvas. In the frightening, enraged tempest, we hauled it over the vehicle. Like tying a handkerchief across your face, the fabric protected the Land Rover's insides, the air filter, and fuel and brake lines from fine dust and sand as well as preventing the interior from being painted in powdery gray. The brim of Patrick's hat swept upward like a sail on a boat and with one hand he pulled at his chinstrap while the other heaved at a rope looped around a bumper. My sunglasses twisted off my face. We were spitting sand.

"Bring the water bottles!" I mouthed, pointing to the back door while standing on the wide rear bumper and gripping a corner of the billowing canvas with one hand. The sand pelted my face as I stretched up to grope for the shovels on the roof rack with the flapping tarp pulling me in the other direction. Throwing the spades to the ground and jumping down, we roped the reluctant last corner of the cover to the bumper. Pausing to look, we realized there was no question about staying inside the Land Rover and eventually falling asleep. The doors could be blocked with drifting sand, and it would become our tomb.

Bent over and bulldozing our way through surging sand, we hobbled for the black breach. It had been shaped when massive pieces of rock had fallen against one another, leaving a space within. Over thousands of years, additional rocks had tumbled from above, sealing and enlarging what was probably an elongated cavity. We squeezed through, away from the rage. It whistled hysterically as if calling us back, gusts of hot, prickling sand arms trying to find us in the dark.

"Did you bring the flashlight?" Patrick asked.

"No . . ."

"Oh shit . . . something else could be in here."

"Matches . . ." I jiggled a plastic vial I carried. Immobile as possible and concentrating on the darkness, I struck one. A whiff of sulfur and the match flickered. Nothing moved in front of us—we inched hesitantly away from the crack. Patrick lit another. The glows ranged surprisingly far in the pitch black. We were in a room-sized cavern and there, in the middle, were the silver and black ash remains of many old campfires. A refuge that had protected Bushmen travelers for perhaps thousands of years! They were the only people in the huge region capable of living under such conditions. "Whew," I whispered, "we got lucky."

"It's relative at the moment; have to check everywhere," he replied.

The worsening storm quickly darkened the entrance. Patrick moved forward, another match held in front of him, peering around the sand floor, looking for things like snakes and scorpions seeking shelter as we were. Something could be silently sulking in a corner, like a hyena or cat; but it seemed likely they would have already issued a warning. We searched carefully. Only blurred remains of human and animal prints lay in the packed sand, their presence recorded as if in a book. Relieved, Patrick began looking at the walls more closely. On a large, flat surface, there was something, an image of some sort.

We inched closer, still wary, and very wide-eyed as we focused. Suddenly, a painted herd of fleeing elands emerged from the wall. In the simplest, most expressive sweeping lines and bold colors, the animals were at full gallop. Caught in midair, some leaping gracefully, their four hooves pointed backward, their heads held high, and their eyes burning with terror of the small men pursuing them. With buttocks as pronounced as watermelons and erections like upright elephant trunks, the hunters were alive with purpose.

I knew that in many places in southern Africa, hunter-gatherer Bushmen, or San, the world's best hunters, painted with great skill and loving care the animals they pursued. But I had never seen any such paintings.

In some way, the Bushmen saw themselves in the animals they killed. They shared the same freedom of the endless desert and endured the same thirst and starvation. They burned in the same heat and suffered when gritty gusts tore at their wrinkled skin. When they were too old to keep up with their band, they were abandoned and then died.

To the Bushmen, the hunter and hunted were two soldiers who eventually would meet in combat, and one must die so the other can survive. From these deep emotional connections emerged the Bushmen's fervent desire to pay tribute, and perhaps a debt, to the hunted animal.

Heat driven by the tempest accumulated in our shelter as night approached. We lay on the cave floor beyond the reach of the groping sand that struggled to fill our refuge. The storm pounded and swirled, raced through rock cavities, whistling to trumpet its presence and torment our sleep, until sometime near daybreak a blade of church-like sunlight thrust its way inside. It was magic.

The sensation wrought then by the desert's sudden, utter stillness was otherworldly. We hesitated for a moment as if a heavy weight had been lifted.

Squeezing out of the crack, stiff and disoriented, we stared. Sand—everything was covered in flowing and graceful drapes of pristine sand. Not one blemish.

"Seems a shame to pee on it, Patrick," I commented, unzipping myself.

"Maybe something will grow," he replied, adjusting himself and then bending backward to stretch and spit.

I instinctively kicked some sand in front of me. "Let's check the Rover."

Shovels on our shoulders, we slid forward on fresh, unpacked sand.

"There she is, Patrick, all wrapped up and trailing sand like a bride's dress."

"It will start," he stated, reading my thoughts. We swept sand off the canvas and found the Rover wasn't buried too deep. "If we have to, we'll use the manual crank to turn the engine over."

Lifting the canvas from the door, Patrick squeezed his way inside, where everything was covered with a layer of golden dust. Sneezing and scratching, he pumped the gas pedal a few times. "It's in my pants," he complained.

"Keep it there."

Looking at me, he smiled, raising his black eyebrows, and turned the ignition key. A throaty gasp penetrated the dead silence. It started. He manually engaged four-wheel drive to provide maximum traction on the air-filled, loose, and slippery sand.

Patrick slowly began to rock the Land Rover back and forth as I shoveled furiously. Finally, we backed out. The Rover, still covered in canvas and a blanket of sand, looked like a weird float in a parade.

After sweeping off more sand and removing the dragging canvas, we celebrated with a breakfast of canned tuna on stiff bread. It was about seven in the morning when we pulled away.

Patrick shifted the gears. "What did you do with the tuna cans?"

"Threw them into the cave," I answered, opening the forward air vents and lighting a cigarette.

"Good. I don't like the smell of decaying fish."

Several hours later the few landmarks we searched for still had not appeared. The sandstorm had transformed the desert landscape. I got in the driver's seat and drove to the crest of an enormous dune while Patrick stayed below, making coffee on our gas burner. Using binoculars, I looked for some familiar object on the horizon. The desert sky was cloudless, the air clean and fresh. There was no pollution, nothing. I could see a vast distance. No signs of life, just seamless sand. I was awed and somewhat frightened by the enormity of what surrounded us.

Patrick had spread our crude, partially self-made map on the ground as the water heated. He waved to get my attention. I drove back down and joined him. As we drank black coffee, he pointed to red threads protruding from the sand next to his boots. It was an odd discovery. A moment later and on his knees, he gently scooped layers of sand away from a piece of fabric. Curious, ignoring the heat, we dug, carefully shifting the sand with shovels.

"What *could* be here?" I said aloud.

Patrick paused, wiping his face, and smiled. "I'll bet the Bushmen will be happy to find your tuna cans."

"Yeah, it will be like treasure to them." They would be able to see themselves in the shining bottom and make tiny arrowheads from the metal.

"Did you leave anything else?"

"A pile of crap."

"Good thinking! They will know the gift isn't from the gods and won't be afraid."

Patrick and I continued digging, more slowly now, and passed the canteen between us. We were thankful when the sun finally lowered enough for the Rover to create a shadow.

From the brim of the wide, ever-deepening hole with sand running down its cracker-dry sides, I looked down at Patrick as he spat sand out from under the bandana around his face. I had mine on, too. *Desert bandits,* I thought again. Yet here in the desert, as at sea, concepts like theft changed with the wind, like the massive sand dune above us.

The bones of an undersized hand appeared. The body of a small man followed. The dry air had preserved him. His skull was strangely flattened and distorted from the weight of the drifting sands. His bones were broken. His teeth were worn—indicating he'd been old. He may have been buried here for a hundred years.

He lay in a crawling position, one arm partially reaching forward, where he died. Nestled with the body, we found a charred tortoiseshell, bits of an ostrich egg, a small bow, and a few arrows that crumbled under our touch. In silence we swept sand away to reveal the tips of seven horn paint pots, which had hung from a crumbling thong belt. Each pot contained the dried remains of a different color. He was a Bushman; his size, distinctive bow, and frail, three-part arrows identified him. And he was one of many Bushmen artists.

We found a fragile leather pouch. Parting the dehydrated skin, we squinted at its contents. Colored, smooth pebbles and pea-sized pieces of what looked like broken glass fell out.

"I'll be damned!" Patrick whispered as he stared at the transparent pieces he had separated out and placed in his palm.

"Diamond?"

"I'm not sure, but it's easily possible." He held the largest one up between thumb and forefinger then looked closely at it with his hand lens in the sunlight. "It looks like a partial octahedron. It's symmetrical, its planes catch and reflect the sunlight, but there are fracture lines.

An industrial-grade diamond? Not a valuable one, but a diamond? Some seem to have paint on them," he continued, looking at the others. "Anyway, *we* can't find diamonds here, can we?"

"No, it's illegal. But why would *he* want them?"

"That is the question. Let's make camp. We can talk about this and make a plan. Tomorrow morning we need to move on. If the guards find us here, we'll be in jail, after they and their dogs jump us."

At dusk the desert became cool and the stars so clear that every mythological configuration in the sky seemed very close to us. Our fire, made with wood we carried on our roof rack, was nestled against the steep side of the dune I had driven up hours ago. We sat with light jackets on, watching and sipping coffee as it cast a restless half-circle of light and warmth over us. Above, a single row of sand grains formed the knife-edge crest of a beautifully sculptured wave. At times the uncanny stillness was penetrated by a slight whisper of wind, immediately followed by a powerful, explosive gust that tore at the crest. The sand grains shot forward, tumbling in one violent yet predictable instant. With this, the smoking sand dune grew and moved imperceptibly forward with timeless dignity. Patrick and I, thoughtful of our day's activities, gazed at our campfire and talked long into the night about the strange Bushman grave and his collection of stones and about the origin of the diamonds buried in the sands surrounding us.

Our evening story began tens of millions of years ago. Then, to the southeast of us and at least five hundred miles inland from the coast, in South Africa, diamonds formed. Diamonds are created in the earth's mantle under tremendous pressure. They are transported to the surface by deep-source fiery volcanic eruptions that often form cylindrical kimberlite rock pipes, also called blue ground. The blue ground is identified by diamond prospectors. Over the ages these rocks were eroded by wind and rain. Their contents, including the extremely hard diamonds, were carried along a prehistoric drainage network of rivers, including the current Orange River, all the way to the Atlantic Ocean. There the diamonds were carried north by strong long-shore drift produced by powerful prevailing coastal winds and the cold Benguela Current. These forces deposited diamonds sporadically along the desert shore of the African coast. Over millions of years the waters gradually receded from the land, leaving that ancient shoreline where we now sat, some fifty miles inland from the Atlantic Ocean.

Winds and shifting sand reconcentrated the inland diamonds, causing them to lodge in the exposed cracks and fissures of the solid rock beneath the sand. Ancient streams then sporadically carried the diamonds back toward the receding ocean. In the early 1900s, diamonds were found scattered throughout these inland expanses all the way to the coast. Mining rights were later granted by the South African government to De Beers; they sealed off the zone from the public and hired hundreds of black men to collect the plentiful diamonds.

Wandering about, our Bushman might have come across men gathering the small, hard shards that looked like glass. Muffled against the sand-laden wind and slowly moving about on their hands and knees in a line across the valley floor, these men, likely neighboring pastoral Khoikhoi (or "Hottentots," as the early Dutch settlers called them) who had migrated northward to the region, would have looked very peculiar to a nomadic Bushman.

Hanging around each man's neck as he crawled along and swinging with the slow rhythmic movements of each limb was a container like a jam tin. In this the Khoikhoi dropped the diamonds they had picked up from crevices in the bedrock or from whatever else had caused the diamonds to stop tumbling. For each diamond found, the white overseer who walked behind them paid a few pennies.

Our Bushman, very afraid of both the white and black men, would not have approached them. In those days the Bushman was regarded as nothing more than an animal. Older Bushmen were killed on sight, and the children and young women were used as slaves. (In fact, Bushmen had been chased north since the 1600s and 1700s, when the Europeans first settled the Cape Town area.)

In the early 1930s, neither Khoikhoi nor Bushmen would have understood why the strange, hard little stones were of value. The Bushmen had no practical use for pebbles that were too brittle to be hammered into arrowheads and too hard to be strung into necklaces. Also, they had no desire to accumulate material possessions, for there were no privileges or benefits from such ownership.

Yet, our Bushman seems to have collected the glassy stones and colorful pebbles. Why?

Curious, he probably crept into the area where the Khoikhoi worked. Perhaps he did that as they slept in the bright moonlight that reflected off the sand-white desert floor. His unusually keen eyesight would have picked out the tiny, glasslike objects among the Precambrian rocks and windblown sand. I imagined him later shattering them gently with a rock to expose fresh surfaces and crystal faces that flashed colors at him when held up to the strong African sun.

Inquisitive, as artists are, he would have studied them closely and discovered the light was shattered into the rainbow hues of the spectrum. Red as bright as a severed artery, green as grass after a rain, the violet in a rare thunderstorm. As an artist, he would have been mystified, if not overwhelmed, by it. With this magic, the Bushman was able to compare the richest hues of the firestones to those he coveted for his rock paintings.

We speculated on how he died.

Perhaps the day came when he was too old to keep up with his band and was benevolently left on his own to die. Traditionally he would have received an ostrich eggshell filled with water as a canteen and whatever bits of food his small band could spare. This, together with his bow and arrows, the diamond-bearing pouch, a skin bag containing the eggshell, and his paint pots were all the possessions he had. The withered little man could have hobbled off, wearing some red fabric a Khoikhoi had left unattended. (Bushmen believed no one owned anything.) Deep into the desert, he looked for a place to die with dignity, as he knew his time had come.

What of the charred tortoiseshell we had found with his body? Surely his will to survive made him search for food. He would have been ecstatic to see a row of small, sand-filled indentations partially exposed. He knew from the tracks that the reptile was crawling fast and, like him, must rest soon. A short time later, the hunter-artist likely located the tortoise near the front of a soaring dune.

Our Bushman had now found food. Rotating a stick among a small pile of dry wisps of grass, he created a modest fire. Soon the reptile simmered in its own shell, overturned. It was his last meal. Perhaps as he sat beside the fire, he decided he would never leave the dune. In a night or two the hyenas would find him and rip him to pieces. Thinking of that, he became aware of the soft showers of sand as they welcomed him into the desert's womb. Perhaps they would bury him? Perhaps someone else would come along and bury him?

After drinking and eating from the charred shell, the hunter-artist slowly and painfully climbed the three-hundred-foot-high dune and waited on the crest for the gust of wind to find him. He faced the day of death, the day of the hyena, as he soared off the dune with the stones we let him keep forever.

Chapter 2

1975: Into Apartheid Africa

The sight of corrugated-tin shacks dotting barren, flat, sandy gravel startled me. They stood bundled together in loose groups, separated by barriers of old automobiles, driftwood, doors, broken chairs, and tangled rusty wire, carelessly linked together. Wood-smoke genies puffing from chimneys disappeared like hope. A twirling dust devil leaped from the trodden earth, lifting up a newspaper page a hundred feet. It looked like the flailing wings of a startled seagull. I watched it collapse as quickly as it had formed. Behind it were ungainly outhouses, propped up in trampled areas where a few chickens pecked and a goat scratched. This was the Cape Flats, a dumping ground for apartheid, a home of segregated townships, and an ill-advised attempt to mold the country into divisions probably as secure as the barriers I saw.

We drove slowly along the N-2, Settler's Way. A driver from the University of Cape Town (UCT) had picked me up from the Metropolitan Airport. I was there to study geology in the Precambrian Research Unit (PRU), a part of the UCT geology department. My aim was to obtain the last and highest degree—a doctorate in science, or a PhD. I had four degrees now, two from American universities and two from the University of Helsinki, Finland, where I'd just come from. I was fascinated by knowledge, one of those people who follow their dreams.

"Separation between white and black," my "colored" UCT driver stoically explained to me. He was a descendant of interracial sexual unions between Western European males who coupled with Khoisan, Bantu, and Asians in the early seventeenth century. Rounding off the mix, female slaves from the Dutch East Indies were also brought into the Cape Colony. The resulting people were simply called "colored." I wondered if I would be stepping back into the American South. *Would this experience be like living history?*

My hometown is Buffalo, in western New York. I left it behind when I was seventeen, slipping away into what I believed was a gilded tomorrow—the world of adventure and imagination, of the smell of summer grass and wild things hidden in tall trees and secret burrows. My heroes were men like James Fenimore Cooper, Jules Verne, David Livingstone, Mark Twain, and Sven Hedin. These were the stepping-stones I trod on, the glorious wonders and innocence of youth that had brought me here.

I would soon learn that Africa was a land of unusual extremes with often no middle ground; it could be an ethereal place where life and death rose as naturally as the sun. Now, however, my thoughts were simpler: *What would I find at the university?*

The PRU's director, Dr. Manfred, was in his mid-thirties, tall, handsome, charismatic, and a classic efficient German. He had invited me to join him a year earlier. I was given a separate room to work in and introduced to six other graduate students from Britain, Ireland, Holland, and Rhodesia. Basically everything was new to us and we relied on Manfred for orientation. Unfortunately, he was mostly concerned with his own research and accordingly rather impassive about students. Still, he did provide assistance to the point our British colleague was able, in stiff upper lip, to declare, "We will benefit from whatever he casts off."

I spent several days settling in, which included finding an apartment to live in. My program, I quickly learned, was to be mostly self-education, in that there were no classes to attend. Manfred suggested I investigate metamorphic zonation in a portion of South Africa called Namaqualand and then write a dissertation on the subject. Specifically, I would study a fifty-mile east-west strip of land. It was about half a mile wide, including long stretches of two rivers, the Buffels and Swartlintjies, which appeared as weak, squiggly blue lines on our map. That meant they were dry rivers, probably active only once a year with the spring rains. I pictured them being bone-dry and crawling with snakes, scorpions, and other members of a desiccated community. After all, the area was within the semidesert northwest portion of South Africa called Namaqualand.

When I asked Manfred about snakebite kits and reminded him I would be turning over rocks, he said, "Oh, don't worry about the snakes." His knee-jerk confidence and disregard for obstacles reminded me of a used-car salesman. As long as the buyer had a driver's license and could drive the car out of the lot (and preferably out of sight), it didn't matter if the wheels wobbled. That was Manfred: He made the arrangement, grinned with accomplishment, and then moved on.

The good news was when the rivers flowed each year, they cleared off the windblown sand that covered everything, leaving the underlying billion-year-old Precambrian rock types visible. Those rocks would be the subject of my work. Considering our earth is about five billion years old, the Precambrian period contained some of the oldest rocks visible to man. They were difficult to study, as they had been beaten up quite a bit over time. Still, the violent episodes of wind flagellation, weather shifts, and physical changes over millions of years made them more fascinating.

Precambrian rocks form the strong, stable continental shields we live on. They are the building blocks of our planet. My task would be to study their geochemistry, metamorphic petrology, and structural geology features in my specified area. What had happened to them since they were born? I would add that information to existing worldwide Precambrian-rock family trees. To do that, I'd spend portions of four years living with them in the field. After some time, a relationship would develop. It was a bit like getting to know a woman. You whispered to her, slept with her, licked her, and scrutinized every part of her. Then she talked intimately to you. Rocks did the same, only in silence; you *felt* what they told you. As my rocks were very old, they had many secrets—deep ones.

Before long, and after reading as much as I could about my field area, someone told me, "It's better to get there soon and acclimatize before the appalling summer heat begins." Further comments made me wonder if anyone could get used to temperatures over 100 degrees Fahrenheit. Wearing denim jeans, a blue T-shirt, ankle boots, and carrying a wide-brimmed leather hat, I signed out the new white Land Rover allocated to me. My destination, the town of Springbok, lay about 350 miles due north of Cape Town. I would probably return in two months or so and go out again, rotating as needed over several years until I covered my field area.

* * *

Off I went, towing a small, one-man trailer along the N-7, which was and still is the main north-south road through the western region of South Africa. The little-traveled two-lane way links Afrikaner (essentially white people of Dutch descent speaking Afrikaans-Dutch) towns, which pop up like welcome stepping-stones across emptiness. The flatlands of the Cape stop dead at a steep and rocky escarpment about one hundred miles north of Cape Town. The road struggles and hairpins its way up the long lip of the cliff face. My Land Rover made it to the pinnacle, struggling in the lowest gear like an exhausted runner. Once I gained the top, however, I felt as if I were coming out of the neck of a funnel. The mysterious, barren heartland of South Africa opened up before me, a huge span of emptiness. Stony sand to the right, stony sand to the left, spidery shrubs here and there, and nothing else straight ahead as far as the curve of the earth. Above, the blazing yellow sun of Africa poured down, coloring the sand in its own image. Heat waves rose, creating a hazy, shimmering curtain that was always in the distance.

The road north from the escarpment is flat. It boasts neither trees nor grass, except that clinging to the banks of the spectacular Olifants River, which it parallels to a town called Klawer. Announced by a boisterous concert of birds hidden in its greenery, the Olifants is a fantasy—a gift to the dry land, begun in the nearby Cederberg mountain range.

A few carts, made of simple wooden platforms set on back axles taken from old cars and pulled by nodding donkeys, eased on to a verge as I approached. Mostly they were laden with plump hessian sacks of flour and corn. The drivers were Cape coloreds, hunched over and wearing slouch hats with wide brims, looking so lethargic they could have been deceased. I thought that even if a donkey died in harness, it would take some time for the drivers to react. The air had that fungus whiff and feel of talcum dust and nothing at all. It was Africa.

Near every town, flocks of off-white sheep drifted down the roadway, nudged languidly by their unhurried shepherds. Nibbling at scattered grass blades and things not visible, they never lifted their determined heads from the dust, which rose in small clouds around their pawing hoofs and followed the flock like the tail of a comet.

Farther along, as lonely as a lighthouse, the little town of Van Rhynsdorp (now called Vanrhynsdorp) appeared: a simple, white-washed, one-story concrete-block collection fenced on the roadside by a forlorn row of telephone poles marching off into the distance. In keeping with the overall sluggish impression, the Troe Troe River crawled desperately through the town as if on its sad way to a slow, dry, strangulated death.

However, Van Rhynsdorp is the gateway to Namaqualand. It's where rooibos (bush tea) was discovered in the early 1770s. The little town was once inhabited by the Khoikhoi, some of whom lived in caves in the nearby Gifberg and Matzikamma mountain ranges. The Khoikhoi were driven north from the very southern tip of Africa by white settlers. They, like the equally pursued Bushmen, have myths about their beginnings. Actually, a great number of stories told in Africa relate to people's origins. One in particular was told to me.

It starts like many Bushmen stories: Long ago, when the sun was very small and the earth was young, the Bushmen and all black people lived together. In their village there was a long rope made partially of strong cattle leather and partially of weak twisted plant fibers. The two pieces were tied together with a knot in the center.

One day a Bantu said he would use the rope and picked up the leather end. At the same time a Bushman said he would use the rope and picked up the fiber end. The two men quarreled.

Other Bantu and Bushmen came to help their friends, and soon there was a huge tug-of-war, both sides pulling and struggling. The knot suddenly broke, and the Bantu were thrown backward, rolling and tumbling far away to a country where the grass is green and the cattle grow big and fat.

The Bushmen also rolled backward, in the opposite direction, far to the north, stopping only when they reached dry desert. They live there to this day, hunting and gathering without cattle.

The Bushmen say that if they have a chance to grab the rope again, they must take the leather end.

Once a year, in August or September, spring rains come and the drab fields surrounding Van Rhynsdorp produce a vast sea of wildflowers in an impossible spectrum of colors. After the briefest pollination honeymoon, everything wilts—then goats and sheep feast on what remains.

I drove on to Springbok, a frontier crossroads town with a little central hill from which sand roads diverged. There were no streetlights or trees to define the roads, just packed, gravelly sand. People parked most anywhere and walked along rough sidewalks defined by logic and footprints. There was a preponderance of Land Rovers, jeeps, and pickup trucks—*bakkies*.

A couple of restaurants, a wonderful screen-door butcher shop hissing with the sound of knives being sharpened, a down-to-earth furniture store, and several clothing stores defined the business portion of town, as well as two comparatively large, old hotels for white people. One had a sweeping wooden veranda surrounded by gnarled trees. Old men with granite faces and wearing huge, old brimmed hats sat smoking, coughing, and dozing. Afrikaners, mainly—they didn't laugh much, I learned, unless they drank. Women were not allowed.

As I would later learn, 80 percent of the town's population were honey-skinned Khoikhoi, 15 were percent white Afrikaner, and the rest were blacks of various origin. Although integrated shopping and living was prohibited, all residents walked the common streets together. They diplomatically subscribed to the apartheid system . . . until the sun went down. I discovered nature's inescapable mixing force, cross-boundary sex, occurred as often as obscurity took place. Heavily tinted car windows assisted. The night was slippery with foxes chasing young hens.

A day after I arrived I sat having a sundowner Castle beer on the hotel's veranda. An older man next to me mentioned he was a former hunter from Rhodesia. That meant he was a professional big-game hunter who pursued his occupation in Africa. Most of them come from Europe or North America. He revealed finding gold. "It happened when I was tracking a rogue elephant that had trampled people."

Rhodesia, gold, a rogue elephant . . . ? I stopped breathing for a moment. Could that be? He *did* say "gold"? As I sat openmouthed, the old man showed me the nugget he picked up in a stream as he hurriedly tracked the fast-fleeing elephant. As a geologist, I knew pieces of gold could be found in streams, having been washed down gradient from a larger parent deposit. My imagination followed the bouncing nugget downstream as he spoke.

Two long years later, my colleagues and I, following the old man's pencil-drawn map, drove thousands of miles, crossing the Kalahari Desert, and entered the civil war in Rhodesia in search of the old hunter's gold stream. This was Africa in the seventies and me in my twenties, when such untamed dreams were possible.

However, my immediate task was to talk to people and poke around, looking for Precambrian rocks. In other words, to get to know the area I was to study.

Chapter 3

God Gave Us This Land

Since I'd arrived, I'd heard the Afrikaners were called Voortrekkers and Boers. Some rather derogatorily nicknamed these men "rock spiders." It was essential to learn more, as I would live among them for years and I couldn't collect the rocks I wanted without going on their land. Many, as it turned out, were headstrong, opinionated, sunburned, honorable farmers—and that often led to unusual incidents.

The morning heat had not yet arrived as I drove bouncing westward on a rough sand and gravel road to where I believed an isolated farmhouse was located. I slowed as a rock dassie—a chubby rodent similar to a woodchuck—scurried across the track on his way to gray rocks infrequently protruding like pimples from the sands. There was nothing else around. Intending to ask the farmer's permission to go on his land, I was surprised to see a crowd of people standing around the two- or three-roomed concrete, weather-beaten farmhouse. Something had happened. Instinctively, I felt I should not be there. However, it would probably make it worse if I now turned around to leave. I had been seen. I remembered a few days prior when a young farmer had told me to leave his land and thrown a rock at my feet.

I parked near the house and asked a roughly dressed man wearing a broad-brimmed, sweat-stained hat for permission to go on the land. He disappeared inside and then came back, saying I would have to ask the owner. He gestured toward the door. "My father is inside on his deathbed."

I was shocked. "*His deathbed?* Oh . . . I can't disturb him!"

"He wants to know who's on his land," he replied, staring at me.

I went inside—everyone there was watching me—and approached an old, tired-looking man lying in a metal poster bed.

"I'm a student," I whispered. "I study rocks."

He looked at me for several moments.

"What do you do with rocks?"

"Look at them with a hand lens after I break off their weathered outer exterior to see a fresh surface," I said after a moment of thought. "That helps me name them. Sometimes I sprinkle a little dilute hydrochloric acid on the light-colored ones. If they effervesce, it means they contain carbonate. If my compass goes in shaky circles, I know the rocks are magnetic. If they are important, I hit them with a fourteen-pound sledgehammer and take a piece with me. I'm new here, but after a while I will develop a feeling for the land I study; geologists do that. They become part of it. The rocks silently talk to us."

He stared at me, then smiled. "Soon I will be part of the land. I've talked to it many times." He gathered his strength and continued. "Leave the land as you found it."

"I will, sir."

"I hope your conversations go well." Then he turned to his son and said, "Let the student look and take rocks."

I thanked the farmer then said to his son I would come back in a few days. It was an Afrikaner farm: The land was the only thing there was. I was glad the conversation had gone well. When I returned, I saw the overturned sandy soil he lay under, not far from his now quiet and empty home. A simple concrete tombstone rose up awkwardly. I stood next to it. A few football-sized Precambrian rocks had been inadvertently placed around the base of the chicken wire fence surrounding it along with other rocks. They were what I was looking for. I figured he wouldn't mind if I took them.

Who were these Afrikaners? I wondered as I drove off. Of predominantly Dutch origin, in the 1800s they had singled themselves out from the Dutch and British officials who ran the Cape area by naming themselves Afrikaners. Growing tired of being taxed and lectured to by the officials during those early settlement days, they decided to move farther inland, and in doing so further defined who they were. Using oxen and covered wagons, these white Afrikaner men, women, and children trekked from the Cape into the remote north and northeast interior of southern Africa. There, in the flat, open, empty, dry ranges, they claimed huge plots of land. Families multiplied like mosquitoes by having a dozen or more children. When they lined up for a photograph in age-order, they looked like the edge of a steep stairway. They saw apartheid as the will of God.

The Afrikaners' land was passed down from father to son, and their very survival depended on it. "God gave us this land," they claimed. "It says so in the Bible." They cultivated the concept in church each Sunday as they made themselves feel superior and safe. Their proud and determined nature stood out on their unbending faces; they believed God would take care of them. In the rural places where they lived, anyone who disagreed with their philosophy was basically shunned. As long as there were no external winds to blow on their apartheid cloud, it all held together.

I continued westward in my elongated field area to within thirty miles of the Atlantic coast and reached a point where a very steep, rocky-faced escarpment presented itself. Below that eight-story edge, the barren coastal region began; and within it, miles farther ahead, the bone-dry Swartlintjies riverbed appeared. I would look for rocks throughout this area.

* * *

Standing on the escarpment lip one day toward evening, I stared downward across the forlorn sandy plain. A track headed off into the distance. Heat waves shimmered upward in all directions, curtaining off the horizon. Although it was not termed a desert because rain fell once a year or so, it sure looked like one to me. The sun shrouded me in heat so intense it seemed heavy. I struggled to believe that the cold, distant ocean, far out of sight, actually lay just beyond. Behind me was my Land Rover. Should I face the challenge of descending on the very steep hairpin-turn track, thin and carved from solid rock? Ripping a tire or sliding off seemed all too easy to do.

An intriguing feature I named "The Oasis" captured my attention. Defined by eight palm trees that stretched sixty feet in the sky like a group of sentinels, it had almost monumental status. I saw a small house with a twenty-five-foot-tall native sweet-thorn tree growing in its courtyard. Palm trees were not indigenous, and I hadn't seen any others planted in Namaqualand. They were an odd display of grandeur, encouraging my imagination. Whoever planted them was not typical. Yeah, I had to go forward and see what was there.

Slowly inching down the escarpment in low, whining gear, the box-like Rover's nose fell and the blue of the sky disappeared from my view. I pushed myself with locked arms away from the steering wheel while my notebook and everything on the front seat slid to the floor after the first bounce. Pushing hard on the brake to slow momentum, I looked down directly at the road like an arched-over vulture might look down from a tree, undecided as to what was below. It was scary. Steering is a mental and physical process. Eyes follow the target while arms and hands react based on the knowledge given them. If the two didn't work together, I would go off the thin track and likely plummet to a rolling death.

Almost giddy with relief, I reached the plain and the Rover leveled out. *Hell, that was a pisser . . .*

Coated with wafting sand, the road ahead appeared golden in the blazing afternoon sun. A yellow sand and gravel track leading to the strange palm-treed Namaqualand. It reminded me of Dorothy following the yellow brick road to Oz. I stopped and urinated on withered blades of something brown. Climbing back in and feeling oddly upright, I changed into less noisy gears.

A mile or so later and at a right angle to my sandy road, the track leading to the palms and farmhouse appeared.

It was obvious The Oasis had been vacant for decades; it lay passively decaying. No one watched over it, and there were no tire tracks or signs of life. Next to the palm trees I saw a tall, old-fashioned metal windmill, continuously pumping water into a pool made of odd-shaped, basketball-size rocks cemented together. Some were Precambrian age.

The pool, about four feet deep, served as an irrigation tank. It provided water via a very clever system of gravity-driven seepage channels leading to the furrowed remains of a once-sizable family garden. I guessed even in the hottest summers the rock tank would be full of water. It had a slight odor of sulfur, as it came from a heated portion of the earth far below the surface, making the water warm.

Hot and dirty, I climbed into it as darkness fell. Naked and floating on my back under the superb desert air, the sulfur smelled stronger and prickled my body. I could see far into the sky; there were no industrial-produced particulates to obscure vision and make the southern stars appear to twinkle. A parade of enormously bright planets stared back at me. It felt kind of sacred, a privilege in my life I felt thankful to have. I kept hold of that private world, thinking about my travels, thinking about where I would go as my life passed. Reality came back with hunger.

The floor of the Rover's back door served as my chuck wagon. I pulled out and screwed a cooking platform to a gas cylinder, then carried it and my food box to the shelter of the pool wall. Tonight I would cook my version of mulligan stew.

In the absence of fresh meat, it was one of my favorite meals. Normally it consisted of finely chopped onions, peppers, garlic, a pinch of lemon pepper, and a half-cup of wine. To this I added several cans of chunk tuna fish in oil and stewed tomatoes. It was a very nourishing concoction. Tuna and tomatoes retain their strong taste and nutrition better than most canned food. Due to the ample liquid content, it was very satisfying at the end of a dehydrating day, the kind of meal I looked forward to.

As the stew burbled, its aroma permeated the air.

The evening wind increased and blew above me, over the top of the rock walls that shielded me as I ate. The windmill blades purred softly while the long, stiff palm leaves rustled and brushed against each other. The machine's iron frame creaked, making its own rhythmic sounds as it lifted water with its plunger and poured it into the pool. I lay down by the side of the pool and listened to the peaceful orchestra. All alone in the immense empty darkness, challenged only by the crescent moon above, I curled up in my blanket. It was just enough to shield me from the chill of the night.

I wished I'd had someone with whom I could share moments like this. Yet who would appreciate the isolation and hardships of heat and the uncertainty of being in such an area? No woman I'd met would have endured it. I fell sound asleep without dreaming until morning's sun and the subtle sounds of lone birds and scurrying things woke me.

Morning came. I ate the remainder of my stew with great appreciation and followed the short, sandy walk from the palms toward the abandoned farmhouse. Small, square, and slumped, the farmhouse looked discarded and forlorn, huddled near the ground like something crippled and dead. A thick post supported one wall like an old man with his cane. The straw-and-pole roof had partially caved in. The remaining roof varied from windblown bald patches to tattered, bleached straw sticking up like a haystack. The edge of a glassless window showed walls about a foot thick constructed of rocks fitted in place and cemented together the same way the walls of the pool had been. Thick walls and small windows kept the house cool during the day.

The little two-bedroom farmhouse had one door, which showed no signs of ever having a lock. The owners just walked away, pulling it shut. People of the land did not steal. It creaked as I pushed it open, letting a spotlight of sunlight precede me. A white-faced barn owl fled silently through the tattered roof, its feathered legs dangling beneath it. It left behind small rodent bones, feathers, and scattered balls of excrement.

Inside, I stood, amazed. Painted on the living room wall was an enormous floor-to-ceiling flower. Although crude, it had large bright red and yellow petals, a long vibrant green stem, and three base leaves. Like an apparition, it dominated the roughly plastered white walls that had become brown with age.

How unexpected. Afrikaners normally were rather drab people. It seemed whoever lived here had brightened their lives with the image of a short-lived desert flower. The house was empty now of everything else.

I turned and walked into the courtyard at the front of the house, also defined by rock walls about three feet high, tall enough to prevent young children from climbing over them. The large sweet-thorn tree I'd seen from the top of the escarpment grew in its middle, providing shade. A child's wooden swing hung from wire leads tied to one of its branches. Over the years, as the tree's limbs adjusted and grew, the swing had become lopsided. There was no one to straighten it.

Looking back at the house, I tried to visualize it as it once was when children played. Now it was crumbling, disintegrating. The wooden shutters hung at angles and door paint curled. I searched for graves; there were none. Where were the children now? Had there been enough time for them to grow up? Did they return years later to see where they had lived?

Regardless of the number of times that I drove that road in the following years, I always stopped at that intriguing oasis. It was like a magnet to me, like magic in the desert. *Who would plant palms there? Who would paint a flower?*

Over time I learned that droughts, recorded from the early 1800s and on, overtook this portion of Namaqualand every five to ten years, and some were very severe. I guessed that meant even the once-a-year rain did not arrive. Farmers abandoned their land. Undoubtedly, they resisted until starvation neared. The land was all they had; it was a part of their family. Where do such people go?

I could taste the fine, dry dust on my tongue as I wondered what had driven the owners away.

Years after I first visited this farm, a native Khoikhoi family built a shack near the palm trees. They made a garden in the same place as the original owners' plot and irrigated it from the same windmill and seepage channels. As I looked down from the escarpment, I was shocked to see the palms were gone! They apparently took too much water from the garden, so the new squatters chopped them *all* down. At some point they must have needed a piece of wire; they took one from the swing, leaving the wooden seat dangling vertically.

The farm that had stood under the silver sickle of the moon was no more. I never stopped there again.

Chapter 4

Blue Lace and Hobbit Holes

A month or so after I arrived in Springbok, I first visited George Swanson, an American who was a well-known independent Namaqualand prospector. He was middle-aged, of average size with longish gray-brown hair, and his weathered face matched his battered clothes. His ornaments were a large silver belt buckle holding up his pants and, at times, a bolo tie. Living unceremoniously, he didn't hide that he wanted to make money. To that end, his whimsical smile hid a devious, inoffensive nature; he came across as charismatic. He had claims all over Namaqualand and elsewhere and allowed his findings to be exaggerated considerably if it supported a situation he was involved with. Actually, the same was true about most anything he said. I remember asking him once about the location of Precambrian rocks. "Are they worth any money?" he replied, smiling. I remember George as one of the most colorful and decent men I've ever met.

George decided to make his home and work in Springbok; his parents, working at the local mine, had brought him to South Africa originally. America was too boring for his lively character; he would have no inhibiting boss in Springbok. George, a white man, didn't believe in apartheid; young, pretty Khoikhoi and colored girls drifted in and out of his rural home with their children. It seemed to me a young boy in particular was strikingly similar in looks to George, a point I never mentioned.

Prospecting for minerals included encouraging others to invest in claims George had discovered. The Swanson home, right on the main road from Cape Town to Springbok, stood surrounded by piles of ore from George's various claims, and he showed these to prospective buyers with what seemed to me a disproportionate amount of encouraging details. He had a drilling rig that he dragged around and set up at new claims. From the distant roadside he pointed at the rig with its tall tower and told clients work was progressing. I don't recall seeing much activity, however.

George had a way of getting businesspeople from Cape Town and elsewhere interested in his out-of-the-way deposits. He would agree to meet with them after being certain they had the most positive information based on some kind of high-level gossip network he developed.

"It doesn't hurt to be late for such meetings and elusive with details," he said with sparkling eyes. "It promotes the perception you are in no hurry to sell, and that implies you really have something worth selling. Some initially express great annoyance," he said with a wily grin, "but they warm rapidly when presented with opportunity."

George's mother, who must have been in her mid-seventies, lived with him. "Ma" was a pioneer of sorts, like George. She would spend extended periods watching him, as if waiting for her son to do something that warranted her comment. Then the two would bicker loudly about the improbability of each other's stance.

George appreciated me as someone who went to the field and slept on the ground as geologists sometimes do. He welcomed me into his home, where some of his native assistants ate—like family. He was not normally singled out or commented on by the surrounding community for his illegal association with coloreds. The locals couldn't shun him, as he didn't join their society or go to their church, as far as I could tell. He kind of willingly acted as if shunned himself, while presenting his smiling mask-face and greeting all with resigned pleasantry. In short, he was tolerated but occasionally he overstepped the line.

Once, when George and I sat talking in his living room, he fed one of his several huge dogs—which also had the run of his house—by dumping leftovers directly on the linoleum floor. There was no dog dish. "Why do I need a dish?" he asked rhetorically. "The dog will lick the floor cleaner than it was before. However," he said as he looked up at me, "Ma has slipped from time to time."

One of his dogs, referred to as Rat because he bared his teeth when he defended his food, had a stump tail. The deformity occurred, as the story goes, when he couldn't get inside a car quick enough one dark night as he and George were hastily leaving a local establishment. The tail, caught in the car door, prompted Rat to howl as he pulled frantically inward. Unfortunately, several colored ladies were also in the car. Doors flew open and all of the women ran out in various directions, screaming. Rat yanked himself free (less the very end of his tail) and chased after them. Witnesses impugned George's already fragile reputation, which he could only defend by looks of perfected righteous indignation.

When not eating, Rat lay looking as if he had done something wrong. It was kind of a common expression in George's home.

* * *

One of the more interesting mineral claims George had was a vein deposit of what he identified as blue lace agate. "The blue," as George called it, was a typical agate in that it was composed of mostly silica dioxide, a variety of quartz. The chemical, carried in water of varying geothermal temperatures, crystallized on the sides of rock cavities through which it percolated. In the case of the blue, the stone became beautifully, delicately layered, like lace, in colors ranging from white to blue to purple, depending on the additional chemicals the water also carried. George said it was a unique rock; nothing like it existed anyplace else in the world. While he could stretch the truth, he wasn't stretching it this time.

Although George didn't take many people to his precious stone claims, he asked if I would like to see where "the deposit of blue" originated. At that time he kept the location confidential, as it came from neighboring South West Africa—and perhaps involved tax and security concerns.

"We'll make a stop on the way," he said. I think he just wanted to use my Land Rover; he'd apparently lost his driving license. Also, he was trying to get me interested in his work, and this find was stimulating and real.

We drove a considerable distance to see the deposit. It was exposed by a long trench some twelve feet deep and wide enough for several men to dig in. The edge of the sheet-like agate, about three or four inches thick, projected upward into the trench on a 30-degree angle or so, like a plate. Over the years, George's motley crew of Khoikhoi helpers in their tattered clothes had climbed down into the hole and broken off large chunks of the delicately layered rock, which they handed upward. They slept in lean-to shelters next to the diggings. Most of these harvested rocks were reduced further to inch-size pieces by hitting them with a sledgehammer and then either sold as they were or polished and sold as a rarity with various accompanying stories.

While I was there, George tried to get several international breakfast-cereal people interested in putting a little piece of the blue rock in each cereal box, as they would a toy or some other trinket. It seemed like a good idea, and he bullshitted about the stone's rarity and beauty. However, there could be millions of boxes requiring a piece of the agate, and the cereal people, it seemed, probably didn't think it would be economical.

Another time, unrelated to the cereal effort, George read how NASA's Apollo 17 module pilot, Dr. Harrison Schmitt, described the earth as seen from space. George realized that narrative, and the associated pictures that showed swirling white clouds and blue water, could easily describe his blue lace agate. George pursued the idea by presenting a piece of the blue, polished to the size of a hard ball, to a high-powered ecological fund-raising group in California that had on its board US First Ladies from Eisenhower on, plus the governor of California, Ronald Reagan. The similarities were indeed remarkable, and the blue won several contests from that point on. It became a symbol of the earth to many, and was, George said, featured in both *Time* and *Life* magazines. I don't remember the details, but George did show me a brochure concerning an awards ceremony he and his mother attended. On one such occasion he gave a ring made of blue lace to Clark Gable's glamorous widow, Kay Spreckles, who, being a spirited swinger, suggested to the audience they were now married! It was a brilliant marketing move and orders poured in for the blue. To this day, the blue remains a unique rock that is tied forever to George's name.

As usual with attractive rocks like blue lace agate, people assigned them special qualities. For example, some believed the blue lace could activate inner knowing, perform miracles, and lessen anger, as well as provide confidence and cure nervous speech habits. I often wondered what the half-dressed little Khoikhoi thought in the 90-degree heat as they beat the crap out of that "magical" rock.

One day at dusk after discussing the blue with George—he was always looking for ideas to sell more of it—I met another geologist who worked for him. He introduced me to a beautiful colored girl. She was seventeen and entering her prime—with black hair, lovely strong legs, a youthful figure, and apricot skin. Such a meeting in the 1970s was strictly illegal due to apartheid. Her father was a mine captain, a position of importance, so she attended the local high school for Khoikhoi but could not speak English, only Afrikaans. Her high-school uniform was a short dark dress, white blouse, white socks, and black shoes. I understood she had not been with a man before and, offering her a hard-to-find lollypop, did not tamper with that too much.

* * *

After visiting George for a few days, I towed my one-man trailer to a new location in my field area. In winter when the nights were colder, I slept and cooked inside and then lay reading under a large blanket. A Coleman gas lantern provided ample light. Although small and smelling of sweat, this trailer became home to me.

One night I began reading J.R.R. Tolkien's trilogy, *The Lord of the Rings*. As I read, Namaqualand's vastness and isolation, unconventional vegetation, and strange topography seemed to be applicable to Hobbit adventures. It inspired my imagination as I prowled the barrenness, looking for rocks.

A few days later, when I was quite far to the north of my field area, I saw huge termite nests, some taller than six feet. Cone-shaped, they rose from the sand and bush like miniature cathedrals and were occupied by hundreds of thousands of parishioner termites living very close together and worshipping one queen. The tapering spires, constructed of sand grains and soil glued together with saliva, thrust themselves upward from the flat sand. Baked hard by the sun, they had withstood years of annual heavy rains and winds striking at them. Those that had succumbed looked like the pinnacled ruins of medieval castles. Who could understand how these blind creatures—vulnerable, sensitive to sunlight and temperature, poorly adapted to life on earth—were able to join in an effective whole and become a very successful species on our planet? Their undeveloped individual brains were somehow linked together like separate cells into one working, collective brain that was programmed to construct their homes in much the same way humans work together to build skyscrapers.

As I continued walking in the heat and desolation, I saw that some nests had a six-inch-round hole close to the ground. Made by foraging long-nosed, ungodly-looking ant bears, the holes looked like round Hobbit doors to me. *Could, by some extraordinary chance, Tolkien have gotten the idea of the round door from this?* As I wandered in the heat, I even compared the ugly ant bears to Tolkien's wingless dragons.

Years after this experience I read with great amazement that Tolkien had been born in Bloemfontein, South Africa, a few hundred miles away from where I'd come across those enormous termite nests. He could indeed have seen these nests in southern Africa. I also read he left South Africa for England when he was young, so perhaps my theory was wrong. Still, he could have—after all, look at the story George created out of his blue lace.

Chapter 5

I Know What I'm Getting

Collecting rock specimens for more than a month while talking to people gave me a good idea of my study area as well as the overall geology. I was beginning to know what I was getting into. Now it was time to go just beyond "The Oasis" farm to the dry inland Swartlintjies River, which, according to my map, roughly paralleled the road that led through the sand-covered coastal area all the way to the Atlantic Ocean.

The Namaqualand wind obliterates such remote roads to the point of becoming little more than indistinct elongated depressions, or even nothing at all. During this process, as vehicles pass over them the sand is compacted into washboard-like surfaces at right angles to the track, causing considerable jarring to take place. I had learned from seasoned drivers it was best to speed up and theoretically "skip" from one crest of the corrugations to the next. That's what I was doing now, roaring along the disappearing road and gripping the wheel.

Several times a year, the road has to be maintained and the washboard bumps flattened. Traditionally, a dust-covered enormous yellow grader—the large kind normally used for construction work—is employed for this. Like bottom-feeding carp, the grader's bladelike lower lip pushes slowly forward, leaving a flat, deep ribbon of roadway in the sand.

Late in the day I came up behind one of these graders. Its diminutive colored helper—with his light-brown skin and characteristic almond-shaped face—walked in its wake, roughly dressed with a complementary drooping hat. The assistant was doing his best to keep up with the machine, shoveling away bits of rock and piles of sand misplaced by the grazing blade.

A large wooden compartment, teetering on wheels like an old-time circus wagon or train caboose, was dragged behind the grader. This was the "home" of the white driver.

It was exactly five o'clock. The grader and its straining, creaking caboose abruptly rolled over the sand-track confinement wall it had created and stopped. Often, not even a thorny acacia tree or any other such landmark will influence a worker's decision, just the time.

I knew from past experiences that once parked, the (usually) Afrikaner driver retires to his mobile home to freshen up. Both he and the Khoikhoi have eaten and breathed sand dust all day, so how he cleans up is a mystery. To make the task even more remarkable, the trailing wagon has been engulfed in churned-up sand dust all day; it must have been sucking it in like a vacuum cleaner. Nothing keeps sand dust out.

During refreshment time, the Khoikhoi, without the benefit of washing, cooks a campfire dinner for two. If the wind blows too aggressively, he cooks near, if not under, the bone-dry wooden wagon. This, to my mind, should be of concern to the white man inside.

On that day, the driver waved. He may have seen only a few lone vehicles that whole day, so I decided to ram my Land Rover over the foot-high sand wall. Perhaps he'd seen black rocks somewhere as he drove?

The old, unshaven Afrikaner *baas*[1] dressed in washed-out gray pants and a red shirt greeted me with a handshake while his helper busied himself near the fire.

"*Goeienaand Meenheer*," he greeted me in Afrikaans.

"Good evening to you," I responded as he took his meerschaum pipe out of his mouth. "My name is George. May I join your campfire?"

"*Ach, ja.* I'm Heinrich. I can't offer you much; my boy is making *biltong* now."

"Biltong?" I replied, knowing it was hard, dried meat like jerky. "Did you shoot something?"

"No, but we found a young springbok that had just run into a cattle-wire fence. The sun must have blinded him. The wire broke his neck. Joseph here"—he referred to his helper, who looked to be in his mid-thirties—"supports his mother, sister, and her two children, as well as his own family. Every once in a while we find some fresh meat, and he makes biltong for them. I keep a bit for myself, too, but I chew slowly."

He said this with a grin that displayed an array of worn teeth, several of which looked wooden and badly fitting.

We chatted a bit more. I asked him about rocks; he said they were only in the riverbed I would find farther west. And then I asked how the biltong was prepared.

"Come and I'll show you, George."

We moved to where Joseph squatted.

He stood. "Hello, *massa*," Joseph said, yanking off his well-worn cap and nodding rapidly while staring at the ground.

---

1 Afrikaans term for *boss* or *master;* used by blacks when referring to white supervisors

I didn't offer my hand, as it was not custom and would have embarrassed everybody.

"Hello," I replied. "It looks good." He had been cutting the leanest and best meat into long, thin strips. All around him on the sand were various containers full of ingredients. I saw a wooden tub containing liquid in which the meat would be marinated. "What's in there?"

"Five pounds of *Swartop slat*—Blackhill salt," Heinrich answered, "with about eight or ten ounces of sugar, a quarter pound of saltpeter, and a quarter pound of ground coriander seed."

"What else, Joseph?" I asked, smiling and wondering about his life.

Joseph reluctantly grinned, clearly not sure if he was supposed to, and pointed to a bottle of vinegar and a bag of pepper. He was shy and not used to speaking to strange white men. He pointed to another collection and said it was a mixture containing leaves and herbs.

Just to the side of the wooden tub was a brown paper bag formed to the shape of a bottle. I assumed it was alcohol—not related to the discussion.

"He rubs the meat with the ingredients, and then the whole lot is marinated in the tub," Heinrich continued. "Every so often he stirs the mixture to make sure each strip is well penetrated. We leave the tub in the wagon two or three days, and it shakes itself as we move along. After that, he washes the meat off in brine, one pound of salt to three gallons of water, and hangs the strips up to dry."

"How can you hang them when you're grading?"

"We string them to the sunless side of the wagon and try to cover them just as we do when we find a springbok carcass. There's not enough circulating air inside." Heinrich shrugged. "Birds can be a problem, though. Several times we've had birds circling the wagon." He stood silent, shaking his head. "Can you imagine us out here alone, pushing sand, and a vulture circling above?"

"Would you like a cold beer?" I offered, hoping to lighten the mood. "I have some in a cooler."

He looked at me with wonder in his weathered face. "*Cold?* Sure! Never get anything cold out here."

Later, we sat by a campfire while Joseph served a large platter of biltong they'd been saving. It was excellent—not too tough, just a bit soft with fat attached for taste. Sunset came and the colors of the land slid into blackness very fast while the fire gave off sharp points of colored light.

"Once a year, I travel to Johannesburg to sell *mealie* sacks to the gold miners."

I nodded, knowing that the black miners made clothes from the corn sacks.

"The great stores and masses of people are frightening," Heinrich continued. "I don't feel comfortable at all there."

"You were born here, in Namaqualand?"

"*Ag, man*, I have its sand in my skin."

"People like to stay where they started. Fifty percent of all Americans spend their entire lives living within fifty miles of where they were born."

"I'm living more than fifty miles from where I was born . . . actually it increases some days as I drive."

"Then you're more of a traveler than most Americans."

He smiled at the thought. "Well, I know what I'm getting here and I prefer it that way."

Later, I took my blanket from the Rover and curled up on the still-warm sand not far from Joseph, while Heinrich retired to his quarters. Life just went on, no matter who you were or what you did. Sometimes the idea was just to pass each day as best as you could.

Morning came and I was off, full of ambition, headed along the sand road to where it, according to my map, paralleled the Swartlintjies River. After some time I stopped and parked, intent to walk the half mile to find it.

The river's name means "black rock" because isolated and unusual black rocks are found within the riverbed. Geologically, they are called metabasites. They would form major pieces of the geology puzzle I was putting together and help me draw a clear picture of what took place over time.

Although infrequent, the metabasites did indeed pop out like prominent blackheads from the river rocks of white granite gneiss. Generally, they were oblong like partially flattened wagon wheels, with their long axis orientated east-west. Together the minerals in the rock types had adjusted to changing metamorphic and deformational conditions deep within the earth a billion years ago, before they both rose to the earth's surface. Both rock types—the granite gneiss and the metabasites—were of Precambrian age, but the metabasite rocks were more important, as their color meant they contained the kind of dark minerals that could provide specific evidence of their origin. The trick was to find enough metabasites sequentially exposed along the entire length of the river so that I could determine and record any changes in their mineralogy and chemistry.

The river itself was flat, bone-dry, and in some places was obscured by windblown sand. I could see where the annual torrential water had woven around and over large, smooth rocks in its path. Now it looked, at maximum, twenty feet wide and was interspersed with grooved channels dotted by spindly brush and nests of interwoven sticks and weeds left behind by swift-moving water.

Most significantly, the water, like a diligent housecleaner, had scoured the underlying rock with the swirling particles of quartz sand it carried. This yearly "sanding" wore away any outer layers of weathered rock and left it free of blemishes.

No matter how many times I looked at the river rocks, their remarkably clear detail always fascinated me. My metabasites were beautiful.

As I walked the river, I stopped and took measurements with my Brunton compass of the strike and dip—the direction rocks were orientated from north (strike) and their angle from the horizontal (dip), respectively—of selected specimens that remained *in situ* since they had been formed. This data—hundreds of measurements—could determine the direction of the last pressure or squeezing deformational event. If such an event occurred, it would have taken place deep within the earth in response to internal movement processes. Following uplift to the earth's surface due to erosion of the overburden, I was now able to observe the deformed rocks.

Often there were several deformational events imposed on one another over the billion-year period since the Precambrian rocks were born. Remnants of the earlier events could coexist with the latest, perhaps stronger, event. Each may have a different measurable strike and dip. I had to look closely for evidence. I recorded this data and the kinds of rocks I saw on a map, which, in the end, would be a geological map of the whole fifty miles of my field area. Basically, it was a question of figuring out what came first and what followed.

I also collected softball- to basketball-size samples from each interesting outcrop of rock by hitting them with the sledgehammer I carried on my shoulder. I took photographs of these samples and included a ruler-scale in the picture, the specimen number, and an arrow pointing to north. I numbered each piece and then recorded it on an aerial photograph to show where it came from. Lastly, I put the samples in respectively numbered plastic bags and carried them to the Land Rover in my backpack. The work had to be slow, careful, and orderly, with no room for error. Once back at the university, I would look at them under a microscope and analyze them.

Rock sampling of this nature could become very rhythmic, seductive to the point it absorbed all concentration. But this was Africa, so it was never good to forget that danger could find you when you least expected it. Once, I had to abandon my backpack full of preciously selected rocks when a leopard followed my every move. Another time, lulled by silence and the hot sun, I tripped over a python. Then there was the cobra that reared up . . . luckily several yards away, but still too close.

Returning toward evening to the Land Rover, I made a flavorsome mulligan stew and scooped out a place to lay my blanket and sleep. The wind made a frail whistling sound as it nudged the small, hard leaves of an acacia tree close to me—the quintessential African tree with its skeletal outline, locally called an umbrella tree. Early moonbeams against a dark sky peeked through the drought-resistant, thorny, dry branches. It fascinated me that the trees' meager foliage was concentrated along the flat-topped crown, while their starved-looking supporting limbs were elbowed and bent, resembling the bony framework of an upturned bird's foot. They looked as if the sky was almost too heavy for them to grow under. Without them the flat open land would appear disrobed. It would not say, "Africa."

Morning came quickly. After finishing the remaining stew plus an instant coffee, I went back to the river. As I did every day, I filled my backpack with samples and, like the cartoon character Obelix, carried them to the Land Rover. By about noon I finished and decided to drive farther on to the ocean. At some point I stopped then climbed onto the Rover's roof to peer through the still-shimmering heat with binoculars. An inconsistent green and brown band of shrubbery lay in the distance. It was a line of acacia trees. From my map I knew that must be the boundary of the coastal settlement Hondeklip Bay—in English, Dog Rock Bay—named for a fifty-foot-tall rock that resembled a sitting dog until its head was struck off by especially fierce lightning. On the shore of the Atlantic Ocean and about fifty miles southwest of Springbok, it would be remarkably cold there in comparison to the hot interior I was coming from.

* * *

In the mid-1970s, the scanty tree boundary I drove to was interrupted by a chicken wire gate across the track. Trees had taken root around now-leaning wooden fence posts. Unlatching the sagging pipe and wire construction from an equally inclined metal support pole required strength. Straining, I had to lift the whole thing up and out of the sand. As I further struggled to drag it open, leaving a deep furrow, I realized there were no other furrows. No trace of other visitors. This had to be the only entrance from the land side, as there were no other roads leading to it. Of course, windblown sand could obscure recent tracks, but I had the feeling there were few anyway. Driving through, I dragged the gate closed and relatched its wire loop behind me. You never leave a closed gate open in Namaqualand, for fear enclosed animals like cows and sheep may escape. Though here, it seemed no one wanted to get in or, for that matter, out. So, why would you need a gate? It seemed a kind of *Alice in Wonderland* question.

A few hundred feet beyond, the track led to a huge rusting anchor propped up so that one of its two flanged arms pointed to the sky, signaling a fishing village. The smell of the ocean filled the air. The land had dramatically changed from bone-dry to water-covered, from intolerable heat to Atlantic Ocean–cold, from cattle to fish. And only a dilapidated chicken wire fence separated the extremes.

I followed the track until it gave way to smooth white and gray rocks weakly speckled with black minerals defining the shoreline. Polished smooth, they lay protruding from the swirling, angry water rising high in waves to form a natural harbor. Seals barked or lay contentedly while seagulls strutted and lifted off; their *cai*-ing filled the air. Waves washed the domelike rocks clean. They were periscope portions of the strong, stable, billion-year-old Precambrian shield rock on which Africa built itself.

Standing on the "dog cliffs," I looked out to the beaten wreck of the *Aristea*, which resembled a small freighter. Only thirty years before, in 1945, she was beached when a massive storm swept her to her death on unforgiving rocks. Shipwrecks were not uncommon; I would see many more in the years following in places like the Wild, Skeletal, and Forbidden Coasts of southern Africa.

Not far away, on a windswept hill with a view of the vast, churning ocean, lay the village graveyard. I walked among the tombstones one day and spoke to another visitor. "The sea claims our sailors and we troll it for its fish. It's the way it is," he said looking at a weathered, tilted tombstone, standing among others. It read:

CAPT. KENNETH DEVENISH PARSONS,

WHO PERISHED WITH ALL HANDS

ON HIS SHIP "TEGWANI"

ON 26TH JUNE 1947,

IN HIS 36TH YEAR.

A long wooden pier penetrated deep into the very restless bay amid a tangled green mass of oscillating, ropelike kelp. The dock wearily but resolutely provided anchorage for a half-dozen aged wooden trawlers that sailed deeply into the sea's body to capture pilchards each day. Processed at the adjacent canning factory, these finger-size sea creatures allowed the settlement to survive for a century.

Reaching the dock, I saw rope necklaces garnishing the uneven pillars jutting out from the churning water's surface. Other ropes slapped rhythmically against wood hulls, adding their dull, wet, lulling sound to the aching groan offered by stretched lines and knots of wood in the crosstrees above. Beneath the waterline, a multitude of barnacles attached themselves to the anchorage supports like medals. It seemed a tribute to the partnership between man and sea.

It also looked like something that belonged to the past.

Opposite the pier and shore were about thirty whitewashed concrete-block homes lining dead-end streets. Lacking further architectural distinction, they were generally placed parallel to the seashore in several unevenly spaced rows. A web of pathways led from the homes to a prominent row of battered concrete outhouses, presumably shared, perched high on rocks at the sea's edge. They provided a spectacular view of the churning, cold sea. I imagined the user sitting in the very brisk air, contemplating life. Frequent storms had blown doors and roofs off, however. Necessary usage during such angry events was no doubt made more exhilarating by flapping toilet paper, some of which clung to the sparse adjacent foliage.

Few people were outside in the middle of the day. Pecking chickens, lurking dogs, and goats with clanking neck bells wandered freely. I stopped at a small whitewashed building, which housed a very tidy restaurant—the only one around. Inside were six tables covered with red-and-white checkered tablecloths. No one else was there. The owner was a very pleasant lady who spoke with an accent. She told me she'd come here years ago with her husband. I asked if she found life difficult. "Nay," she replied, "I accepted it, I dinnae have much in Scotland and I've nay gone back. My husband, a sailor, passed on years ago. Ach, he loved the sea and sound of seagulls. This restaurant is how I support myself. I know what's here; I have nowhere else to go."

Years later, when I looked at my notes, I wondered if Capt. Kenneth Devenish Parsons could have been her husband.

I drove to the gas station that doubled as a general store. Pumping a worn metal handle, an older man dressed in faded jeans and shirt gazed at me guardedly and silently filled my tank. He asked me to help him pump and why I didn't speak Afrikaans. "You should learn it," he added with conviction. Inside the store was a crank-up telephone and penny candy in a large jar. Shovels, enamel plates, bolts of bright cloth, hats, lanterns, rows of tinned food, shoes made from rubber tires (I bought a pair), bars of blue washing soap, tin kettles, and bags of *mealie* meal (corn) and flour were all neatly stacked.

One Sunday I saw people walking to the one-room whites-only church. The restaurant lady had told me that some Sundays a traveling Dutch Reformed *dominee*[2] would provide the service, other times one of his disciples. Given the dangerous occupation of the sailors, "release from earthly bonds" was a common topic. Ladies attended the service carrying large baskets and wearing enormous bonnets and long, grandmother-style dresses that protected them from the sun and the blowing sand. Men, awkward in their best clothes, accompanied them. They shuffled out of the church and down the silent streets ready for life.

Isolated settlements like this started with a few families. Often, such societies chose to keep to themselves. There really didn't seem to be much reason for people to visit. It wasn't on the way to anywhere. The children married each other, and the next generation did the same. Inbreeding could lead to deficiencies. Later, on one occasion I met a young man in the settlement who had married his cousin. I asked him why he hadn't looked elsewhere for a bride.

"Because I know what I'm getting," he replied.

I went back to the settlement each time I visited my field area over the next four years, lugging that gate open and closed and rarely seeing any other tire tracks. I told myself it was to escape the

---

2 A minister of the Dutch Reformed Church

interior's heat, to sit in the chilly ocean's breeze on the periscope Precambrian rock and stare at the restless, clawing ocean.

But more so, I think it fascinated me to return to a place that didn't change, a place that was tranquil and survived without influence from the rest of the world.

Chapter 6

Preparing a Rock, Finding a Woman

Two months of hot, dry living in Namaqualand collecting rocks and creating a geological map was exhausting work. There came a point when I had to stop gathering data and start reviewing what I had. That meant a return to UCT, where I would conduct certain tests and hopefully have a little social life. But how do you test rocks? And how can you meet women if you're in the lab all day?

Rock specimens are examined by cutting off a very thin slice of the rock in question. That slice is then cemented onto the middle of a two-inch-long glass slide. The sample is then sanded down and made so thin light can pass through most of the minerals the rock is composed of. This is called a *thin section*. When viewed through a polarizing microscope and magnified hundreds of times, mineral details emerge, allowing the rock to be correctly categorized. Knowing I was the first person ever to see that tiny bit of nature in these thin sections fascinated me.

The physical or structural conditions under which these metamorphic minerals were formed can also be deduced by viewing them in thin sections. If elongated minerals are orientated (lined up) as opposed to random, it shows they have been subjected to pressure as they lay deep under the earth's surface where massive surrounding movement occurs. It's similar to a group where each person is being pushed in the same direction; they fall lined up and according to the direction and the force of the shove.

As I noted earlier, minerals form under different conditions of temperature and pressure. Some minerals form when it's hot, some when it's cold. The appearance of certain mineral assemblages reveals the approximate temperature conditions that were occurring when the minerals reached a state of stable chemical equilibrium. That is to say, they were content with each other (like happy neighbors) and not reacting to form new ("baby") minerals. Most metamorphic rocks are in equilibrium, as the conditions they were formed under were stable for long periods.

Sometimes there are tiny fragments (inclusions) of older formed minerals enclosed in younger mineral bodies. That speaks of a previous life for those older minerals. It can mean the younger mineral had begun to eat or dissolve the older one. If not for it being inorganic, it might be considered macabre.

We also look at delineations. Were boundaries between different minerals sharp or irregular? If irregular and blending into each other, this may suggest they were reacting or "mating" to create a new mineral made from both of their chemistries. This took place because they were not happy with the physical conditions they had ended up with. The new mineral "baby" would feel more adjusted and stable.

The chemistry of rocks and minerals is determined by sophisticated laboratory equipment, such as the electron microanalyzer and the X-ray fluorescence machine. Data from my rocks, obtained by using these machines, was vital to understanding the area's geology.

My plan was that after four years or so, I would have enough information to prove something insightful to my science and present a new, original thought in order to submit my PhD dissertation.

There were no desktop computers at UCT in the mid-1970s, just a mainframe located in an isolated building. We used punch cards initially to enter the programs required to interpret larger quantities of data. If I lost one card from a deck or (heaven help me) dropped a deck of hundreds of cards, I would be in for days of work to sort out the disorder. Every card had to be perfectly punched and in sequence, or the program would not work. Scientists then sent decks of cards, aggressively bound, to one another via post.

New electronic typewriters were scarce and, as a student, I used a manual typewriter to compile reports. Our photocopy machines were primitive. It was standard procedure to send out recent publications as hard copy to other workers. If I needed research from other authors, I wrote to them via land mail and asked for copies. Fax machines and cell phones were still twenty or so years in the future. Libraries did not normally contain all recent, pertinent information. Everything took time.

After reviewing locations, I selected certain rocks from my collection for chemical analysis. As is still done today, each one had to be broken down into golf ball–size pieces with a sledgehammer and rock splitter. This after I'd broken much larger pieces off from their parent rocks in the field. A double handful of golf ball–size specimens from a single rock were fed into a strong jaw crusher, which was first cleaned with distilled water to lessen the possibility of contamination. The pulverized results went into a decontaminated agate grinder until reduced to a powder. This was then mixed and quartered again and again to get a thimble-size sample representative of the huge rock's chemistry. That specimen was prepared for X-ray diffraction analysis that magically churned out percentages of the elements present.

I primarily analyzed distinctive *metabasite* rock, which originated initially with the melting of rock deep within the earth's surface. After forming, these igneous (basite) rocks, while remaining in their solid state, endured additional extreme heat and pressure a billion years ago by a process called *metamorphism*. I was most excited by the resulting dark minerals formed; the work with thin sections made it clear they would provide specific evidence of the temperature and pressure conditions under which the metabasite had been formed. And the resulting abnormally hard rock made a unique bell-ringing sound when struck with a sledgehammer. I could identify the rock blindfolded just by the sound it made. It was not crushed to powder easily.

However, *metapelite* rocks also were found in my semidesert study area. These were formed underwater and as such were of sedimentary (pelite) origin. Later, they too suffered metamorphism by the same metamorphic event that affected the metabasites. Finding these metapelites meant there once had been a lake or large body of water where I searched. Composing some of these pelite rocks were extraordinarily big, prominent minerals like "round" red garnet, "thin and tall" colorless sillimanite, and blue kyanite, which could easily be seen with the naked eye. These minerals had the ability to grow (or crystallize) in a large and distinguishable characteristic manner, unlike most other minerals, which remained small and without a visible form. They were like weight lifters among children.

* * *

It was a hot day when I first decided to begin breaking my specimens down into smaller pieces using the only way possible—brute force. Just outside the geology building garage was an area designated for this. I took my shirt off and was working the sledgehammer as if on a chain gang. My muscles were hard from fieldwork, and the heat caused veins in my forearms to swell, while my longish, naturally curly hair had lightened to blond from sun exposure.

"Don' hit your fingers, To-zan."

I looked up to see a girl smile. Her brown eyes danced with mirth, her white teeth brilliant against her olive skin.

"It would help if someone held the rocks for me," I replied.

"Use other hand," she said in heavily accented English. "Like in garage."

"What?"

"You university mechanic, helper?"

I laughed. "Why would you think that?"

"You sweaty and your clothes, they are dirty. Maybe you fix my old car?"

Her impression contradicted how I saw myself. I remained silent. She walked away. She wore tight jeans and a T-shirt. Her backside was exquisite, going up and down on either side like a seesaw as she walked. I got a brawny taste of desire in my mouth. *A real Jane*, I realized, unable to look away. *Was she curious about me?*

"What's your name?" I called after her.

"Catina," she tossed over her shoulder.

"Do you want to know my name?"

"It's To-zan," she answered, not looking around.

What a snippy little thing! Still, with such a lovely protruding, proud chest and a trim waist emphasized by her tight jeans, she had a right to be. She was beautiful, a few inches above five feet tall; her crow-black hair was pulled back to the nape of her neck and wandered far down her back, lifting gently with each wisp of wind.

I had to meet her.

The next day, I was surprised to see Catina eating and reading in the geology building's small lunchroom. I walked over and asked, "Do you mind?" as I held my sandwich and gestured at the available seat next to her.

"Whaaat?" she asked, half turned in my direction.

"Chair," I pointed with my sandwich.

Holding her sandwich, she looked up from her book, which I later learned was written in Portuguese.

"Chair." I pointed again with my sandwich.

Still apparently shocked at seeing To-zan next to her, she looked at me and blinked. "You want to chair my sandwich?"

Speechless, I shook my head vigorously and held up my own sandwich as if to say, "I have one."

Oh God, does she really think I want to share her sandwich? Not waiting for an answer, she went back to reading her book. I glanced around and sat down. It seemed that everyone in the room was watching.

More discretion was required. I gently touched her leg with mine under the table.

She recoiled slowly, apparently considering the encounter accidental. The second time I did it (stretching) she looked up at me with her sandwich poised halfway to her lovely open mouth in surprise. I would have eaten what she had partially chewed.

She gobbled her food, stood, and slammed her book shut. I guess I'd violated her view of introduction protocol. As she headed for the door, she looked back at me with annoyance.

I'd made an ass of myself.

About a week later, she and I both attended a department social function, and I took the opportunity to explain I was not a mechanic or hired to break rock, but I was a PhD geology student. She asked why I'd rubbed her leg.

"Mine itched . . ."

She looked at me. "What a cheek," she replied with a half-smile. Her eyes were then full of mischief. She had a sense of humor. My heart skipped a beat.

After a drink or two, Portuguese Catina (the nickname I eventually gave her) explained she was a master's student, working and studying in the geochemistry department in the same building as the PRU. "I born and raised in Mozambique with its sugar-white sand beach. I came here, to university, when I eighteen. My parents sent me here when they returned to their native Portugal." Later I learned her wealthy father lost all seven of his grocery stores, taken over by a new black Marxist anti-colonial government based on ideological fantasy. Almost all the Portuguese population was left with nothing, while fearing for their lives. They watched the graceful avenues of flamboyant jacaranda trees hacked down for firewood, while locals moved into unfinished homes and hotels. University students from Rhodesia warned that their country, too, would likely fall within several years. A cloud of fear and uncertainty hung over South Africa, the country bordered by *both* Rhodesia and Mozambique.

With little money for her education, Catina was on her own. It was a story I was to hear many times during this very restless and scary period in southern Africa's history. I liked Catina. She wasn't afraid to go forward with little support in a new country and try to further her education.

We met frequently after that and from time to time discussed her life. She performed research work for others and sent money to her needy parents. Such a reversal of what might be expected of a student. I tried to console her but wasn't able to say much. Not to mention, I found my own concerns overwhelming at times. One stormy evening I found myself in her apartment, lying on a rug and trying to get my large hand up the leg of her tight jeans.

As I was praying, "Please, just a little farther . . . ," a bolt of lightning ripped across the turbulent sky with ear-shattering thunder. Startled, she flung herself at me. Her supple arms found my neck and her legs my waist while our lips became ziplocked together. A moment later she jumped onto the bed and stood looking at me. My eyes opened to dish size; my blood became a raging spring stream. I gasped . . . a prayer answered before my eyes. The magnificence of youth slips away each day, but some magical moments remain embedded in memory. This was one . . . thank you, thank you . . .

Morning came. Sunlight marched into the room. She had risen early and came to me, lowered her hand for a kiss, and then left for the university with unparalleled dignity. I stumbled home wondering if I was still on earth, if it had really happened. It had. Afterward, we became very good friends.

* * *

Several weeks after meeting Catina, I began to realize data from my field trip suggested a story drastically different from what a previous researcher, who had worked in my area, had published. I was flabbergasted.

This researcher claimed that the length of the fifty-mile area I was studying had been "divided up" into sequential metamorphic zones. Each approximately fifteen-mile-wide zone merged into the next, and each was created under definably and vastly *different* temperature and pressure. Those divided zones and the relationship between them were what I was to study. However, I had been unable to identify distinct zones. Instead, my tests showed that all the metabasite dark minerals in each zone were the same; in fact, there were no zones!

My initial deductions said that the whole area was formed under *one* continuous temperature and pressure zone. If I was correct, the researcher, who had received his PhD for his work, had made a huge mistake. In short, he presented the whole region known as Namaqualand in an erroneous geological fashion. How the hell was I to investigate zones that were not even there?

Worse, his published findings had misled all subsequent researchers who built on them. It was a house of cards ready to fall. And now I, a student, was the first to reinvestigate his findings. Would I bring the house down?

"How could this be?" I lamented to Catina over coffee one afternoon.

"You need go back to field and check everything," she advised, looking at me strangely.

I could tell she thought I was wrong. "Catina, it's not unusual for scientists to disagree, but this seems very clear-cut. The previous investigator is an Afrikaner, though, and pretty well known. His name is Joost, and I understand he, along with most Afrikaners, likes to be right. Actually, they are *sure* they are right—and they don't give up. He works for a huge international mining company and submitted his findings as his PhD dissertation. I hope this won't create a problem for me."

Now she looked anxious. She knew what power the Afrikaners had. She knew what could happen when people had power. She would never forget Mozambique. "You probably wrong. You need to calm down."

"Okay, perhaps . . . But his work . . . " I couldn't let it go. "This is tantamount to saying something like two plus two equals five. It's just wrong. The results are a research disaster, like having a basketball team play basketball with an oblong American football, it's that crazy. I can't believe it, or . . . that I'm even embroiled in this."

The whole thing felt heretical. "This fellow has many influential friends," I continued as I rolled things over in my head, trying to understand. "There is an old-boys' network among Afrikaners. They will be on his side. With those kinds of connections, would he let a student—a foreigner—criticize him? Prove him wrong?"

I don't think so. I'm in deep shit.

"Why would he make untruth like that?" Catina asked. "What he gain?"

"I don't know . . . I don't know," I repeated quietly. "Maybe something strange? Whatever, it will have to come out in the months that follow."

"Two and two not five, Beeg George." (That was the nickname Catina had given me.) "You can't make it five, but maybe they won't let you say it four, either. This is Africa. There is no middle in Africa. This university awarded him the degree. They will not want you to show they were irresponsible to accept such erroneous work. Rocks can drop down well with no splash."

Rocks can drop down well?

Chapter 7

"The Last Bushman Was Shot by My Father"

I didn't want to be in the situation I was in, afraid of what I might find. I hoped I couldn't be right about Afrikaner Joost, as he was referred to by some. I said nothing to anyone and packed my Land Rover for the trip north to my field area. I'd been in Cape Town several months. On this trip I wanted to visit a large Afrikaner cattle and sheep farm to the far east of my study area to compare the rocks there with those in the central and western areas I'd been reviewing. Maybe I missed something. I also wanted to experience how rural apartheid, or the master-slave arrangement as it was guardedly called, worked.

The farm was some sixty miles inland from the Atlantic Ocean, in a land curiously pierced by smooth, dome-shaped, pinkish granite-gneiss rock some fifty to one hundred feet high. These polyps resembled massive heat-cracked boils on stretched skin. Due to erosion, they became decidedly convex as the outer layers peeled off, or exfoliated, like the skin of an onion.

In some areas the domes appeared to have comfortably spaced themselves across a sea of sandy, brownish soil. In other areas, they overlapped and formed extended knotted ridges that scarred the perfect horizon like an old, badly healed wound. Vast cattle farms dominated the land, and the dry Buffels River ran through it. Not a cloud was in sight; the sky was infinite blue as I plodded along small tracks interrupted by cattle fences.

The sun beat down. A troop of baboons barked fiercely. I was in their space. Although a quarter mile away, I could see their silhouettes clearly, in single file like slow-moving black roller-coaster cars following the outline of the dome ridges.

I stopped to open a cattle fence, one of many. Next to the gate were the small, dehydrated bodies of skinned Karakul sheep. It was not unusual for farmers to leave such a warning at fence gates, staking their territory.

I looked up at my watchers. They barked an alert. A huge gray patriarch stayed behind the curious foraging troops, protecting their rear. He paced back and forth as if to challenge me. Adolescent males led the group and in the center were the females and their young. Some of the young clung upside down below their mothers' furred bellies, their strength already remarkable. Others rode on their mothers' backs, surveying all with the clear, large eyes distinctive of their species while picking lice and whatnot from their fur and eating them.

They were arrogant, invincible, and ready to fight as they knuckled their brawny forelimbs forward. Their forearms were far stronger than a man's. They could rip most dogs to pieces. Their fangs were several inches long, worse than vampire teeth, and their eye-popping rage, especially to protect their families, unusually explosive. A baboon fights by sinking his fangs into his opponent and pushing away with his well-suited forearms, tearing out chunks of flesh in the process.

The bull stopped and sat his muscled, hairless backside on the very highest curve of the dome and watched me at the gate, realizing I'd not moved. His elbows rested on his knees. He uttered a deep, strong bark—*aaaaaaa-raaaaaaa-hoon*—that resounded easily to where I stood. The call was not savage or menacing. If I'd been a lone woman, he likely would have barked more loudly and made indecent gestures. He was simply asking who I was and warning me to be alert to his presence. I cupped my hands and, in a deep, strong voice, mimicked him—*aaaaaaa-raaaaaaa-hoon*—giving him the same message back.

The patriarch listened. *What kind of bark was that?* he must have wondered. He gawked hesitatingly, his doglike muzzle pointing at me as he cocked his head from side to side. He stood and knuckled his way a few paces in my direction before barking again. I did the same. No threat, the old fellow must have rightly assumed. It's just another asshole human.

The troop moved on, as I did in my Land Rover, remembering one of the Egyptian gods was depicted with a baboon's face.

After passing through seven gates, I reached the Afrikaner cattle farm belonging to farmer James. I drove to the tidy, large, ranch-style farmhouse surrounded by shade trees. By Namaqualand standards, it demonstrated success and wealth. James was in his big barn, wearing a blood-splattered apron over his field clothes. He was treating Karakul skins, which he sold to European merchants.

He explained that Karakul lambs had to be killed within days of their birth, when their skin was jet-black and curling. The resulting fur was in fashion then. I preferred to think those who wore the fur were not aware it was from two- to three-day-old lambs still sleeping in the manger.

Stopping to breathe deeply, he explained he also raised cows. "A few years ago one of them trapped me against a fence and punctured my lung, so I only have one left."

I grew to know Mr. James and his family quite well over the four years I visited Namaqualand. I liked them. They personified the honesty of the land and were very friendly. Perhaps that emanated from having a good and comparably wealthy life, as I was to see. Someone from town who knew James said he had a *very* violent temper. I never saw it, but I never gave him reason to provide an example. Successful at farming and guided by his wife, he brought from Cape Town lurid bougainvillea, heady jasmine, trees, flowers, and other shrubs to embrace the sprawling farmhouse, green lawn, and tennis court that he built himself. There were also fruit trees, which the baboons would strip of all nourishment if given the chance. From time to time the primates would kill a lamb or sheep. The farmers shot them when they could. No one questioned that remedy.

The trees and plants were watered faithfully from the farm's deep wells, but due to the salt content of the water, most died after about ten to fifteen years and had to be replaced. There were no palm trees like The Oasis farmers had planted, however.

The James farm did have a sweet-thorn tree like the one at The Oasis farm, and it captivated me. Also known as *soet doringboom*, the sweet-thorn tree is noted in the light, swingy South African folk-song "*Sarie Marais.*" The song is played by the South African Army, the French Foreign Legion, the British Royal Marines, and others—often during parades. It's possible it's an earlier version of the old American Civil War folk song "Foggy Dew," with words translated into Afrikaans. I found the Civil War connection intriguing. In any event, the tree is capable of sending taproots down more than 180 feet to find water, which allows it to survive in the inhospitable, dry environment without irrigation. Namaqualand pioneers resourcefully used its sweet gum for confections, its wood for wagon parts, its bark for rope, and its leaves and pods for fodder.

An old cast-iron bell hung from one of its limbs. When struck with its clapper, the bell delivered a resounding ring that called the farm's Khoikhoi servants to work. Effectively, it was a slave bell.

Just outside the general area of the farmhouse and within easy earshot of the bell was a row of small, identical, one-room, whitewashed concrete blockhouses. Each one possessed a soot-stained chimney, which led to an inside fireplace where the Khoikhoi residents prepared their evening meals and morning coffee.

The colored women worked in the farmhouse kitchen, cleaned the home each day, changed the bedding, and performed whatever chores the farmer's wife and daughters asked of them. The men toiled in the field, repairing and building fences, clearing planting areas, feeding the farm animals, and herding the cattle from field to field. They worked under the orders of the farm's owner, whom they called *maaster* or *baas*—boss.

"On the weekend," James said, "they get horribly drunk and fight. One man went blind from drinking pure alcohol or something close to it a few weeks ago."

One day, like many days, the earth burned with heat. James and I stood talking under a jacaranda tree next to his home. We watched a colored family in a horse-pulled cart filled with belongings unhurriedly come down the sand track leading to James's farm. It stopped a distance away. The father got out and approached us. His wife and children remained quiet in their wagon, sitting in the sun.

"Maaster, I want work."

"Would you and your family like water to drink?" James replied immediately, without offering his hand. It was custom to not shake hands with nonwhites; indeed the Khoikhoi would have been embarrassed had James offered.

"Yes, maaster."

James had one of his colored workers fetch plenty of water.

"Where have you come from?" James asked.

"Highveld farm, maaster, very far north."

"Why did you leave?"

"Maaster sold farm, and new baas brought his own colored boys with him. Maaster say we had to go but gave food. I work for maaster twenty years, maaster. Food now gone, maaster."

After some discussion, James said they could stay and work while living from their wagon. In return they would be given food, clothing, and enough money to buy essentials. In due course, if the symbiotic relationship worked, the farm community would build them a white concrete house with a chimney.

Few considerations are attached to such a deal. For James the transaction was like buying a horse: He needed to feed it and keep it somewhere. The system depended on the laborer not being able or wanting to do much else.

I assumed the colored people were not educated and asked James about that. He said, "We try with the children, but the adults don't want to be educated. They burn schoolhouses down."

One morning I was in a remote part of James's land not far from a track. One of James's colored boys passed by, driving a tractor that pulled a wagon full of the farm's colored children. It was the daily trip to a one-room concrete-block schoolhouse, just as James had mentioned. In the late afternoon they would be driven back to the farm. Providing this service was one of James's duties.

White children from the farms went to school in Springbok. There they boarded during the week and returned for the weekend.

* * *

Over a few campfires, I thought about James's relationship with his colored servants. Basically, it could be called a form of slavery, although they were free to leave. But, if they did, what would they do? I didn't think any other farmer would take them unless James approved.

The man and his family seemed very relieved to find a place to work and live in peace. They expected the new maaster would feed and protect them. When they died, they would be buried on the farm they had come to know. It seemed a safe path through life, but was it right?

How did this system of governance compare with others like it? I'd read about the Bolshevik form of socialism, for example, that began in the 1930s with forced collective farms in Russia. They did not promote learning; they wanted the focus on physical labor to the point that they killed off the well-educated. I knew firsthand about ownership and duties in the Israeli kibbutz, having worked on one in the early 1970s. The work was shared between all members; everyone worked for the basics in life, living and eating together as equals. If some were not happy with the system, they could go elsewhere without a problem.

At face value, it seemed to me that if people agreed willingly on a lifestyle direction while they had other options to consider (and were able to follow them), then why not? In truth, the deciding factors were far more complicated—and often led to horrific civil wars.

Near the farmhouse, en route to the deeper portions of the farm where I would camp that night, lay the family's cemetery. There James's mother and father were buried among other members of their ancestry. It had a well-kept white wooden picket fence around it and was sheltered by acacia trees. In the near distance and on either side stood two rock domes, providing a magnificent background. Two breasts of the heartland, between which the sand track I followed carefully felt its way. The sand draped off the domes like a long wedding dress and gently touched the carefully placed tombstones. All whites stayed together on an Afrikaner farm, whether dead or alive.

That peaceful presentation did not apply to wandering Bushmen, however. James had told me without any embarrassment, "The last Bushman in the region was hunted down and shot by my father."

That would have been some sixty years earlier, in the early 1900s. Maybe James inherited his supposed violent temper from his father? I mean, who would shoot a Bushman? Still, it was different then; life was polarized into firm viewpoints. The little Bushman, like all

Bushmen, believed that every animal on earth belonged to everyone. That was his system of government. The concept of ownership was unknown to them. The Afrikaner owned his land and livestock and loved them passionately. When a Bushman saw one of the farmer's cows and he was hungry, he killed it.

The Afrikaner farmers honestly believed that the Bushmen were animals, like the hyena and jackal. Sometimes when captured, Bushmen were put in holes in the ground and told to dig for metals. They passed up baskets full of whatever they were digging. Food was dropped down to them. They never saw the sun again and eventually died underground.

When the sun set that evening, I was several miles into an isolated part of the farm characterized by an especially impressive cluster of rock domes that were millions of years old, which I can visualize to this day. They made me think of the remarkable Bushmen who once likely traveled over them, living their ancient customs. Human beings that ended up basically being buried alive. It was a good place to camp, have a fire, and think about life.

Mulligan stew simmered in its blackened pot above glowing coals in front of me as I sat on a rock. The air cooled a bit as a thin column of wispy, warm gray smoke rose and slipped away into the sky like a ghost on a mission. Sundown came fast and brought stillness. Under a desert sky—clear, timeless, and appearing very close—I felt I was at one, lifted up and part of the enormity around me. With only the stars, moon, and sand for company, I thought again how far I was from my hometown and the numerous cabins I'd built throughout my youth and slept in during isolated nights. Those were happy, peaceful days.

Curling up in the sand, I slept and dreamed of the Bushman James's father had killed. With the arrival of the white man to southern Africa, the Bushmen fled north, leaving only isolated bands behind. Given James's father's statement, he must have been the last member of the final fleeing band. Possible he was old and was left behind to die as Bushmen custom dictates. Still, he would have run on his old legs as far as he could rather than give in to the white man, whom he greatly feared.

In the morning, I had a large bowl of muesli, powdered milk, and honey for breakfast, along with several slices of bread spread with large amounts of jam.

Later, while slowly driving over the hard ripples of sand that had formed on my track, I saw two small, flat pieces of propped-up rock on the verge, carried from a distant exfoliating dome. I stopped to stare.

Each foot-high piece stood on its edge a yard apart from the other. They had begun to lean. I saw no particular landmarks around, like a home or building—just sand and the odd aloe tree. They had been placed there, in the manner local natives were often buried, with no particular place to go. The flat rocks were likely the isolated graves of two of the farm's coloreds, probably husband and wife. I suspected they had worked for James's mother and father. In time the rocks would fall over and be covered by sand. Nothing else would record their existence except memory. Their lives were over. They probably hadn't made it past forty years old. I doubted if anyone could recall their names now; the couple had likely worked until they dropped, and then others stepped around them. In my mind it seemed such a system couldn't stimulate thoughts of the future or promote knowledge. *What was life without such mental pursuits?*

Perhaps because I was a young man full of energy, of life, of desire to be all I could be, I found it difficult imagining the constraints I saw. The system I was born under enabled and encouraged me to develop. Apartheid did not.

That's part of what was wrong with it. And the solution wasn't simple at all.

Chapter 8

Oven at High Noon

The night wind decreased to a breath. I kept the Land Rover windows open, as it was still cool and refreshing. That feeling would soon drift away with the swiftness of the African dawn. The summer day would soon quiver with heat.

I drove into Springbok to spend an evening with the old man who'd told me months ago about the gold nugget he'd found in Rhodesia while hunting elephant. As we drank Castle Lager beer on the veranda of the Springbok Hotel, he pulled the nugget out of his pocket and retold stories. Later, driving off captivated and a little drunk, I stopped a few miles outside of town to sleep. The next morning, I drove west along the sand road toward the coast, looking for a place to begin the day's work. I wanted to walk inland to a portion of the lengthy Swartlintjies River I hadn't seen before. A lonely group of wind-tortured acacia trees appeared as I drove, throwing dust and sand behind me. Might as well park there, I decided. Slipping on my empty backpack and hefting the sledgehammer onto my right shoulder, I checked my compass and pointed myself in the direction of the river. It was seven o'clock. I was still wearing clothes from the evening before: cutoff jeans, a loose, white, short-sleeved shirt, broad hat, and ankle-high boots. I began to walk, not considering that I might still be a little drunk. Unfortunately, where I parked on the side of the road, the river turned out to be much farther inland than I thought.

It's a moonlike expanse of bleary sand and skeletal shrub, I thought, finally reaching the riverbed. Peering downward and turning my head like a rooting chicken, I concentrated on locating black metabasite and rarer metapelite rocks. Finding them as sequentially as possible along the entire length of the river continued to be my objective. Comparing them would then reveal if changing conditions of temperature and pressure had occurred successively when they formed.

After searching the river's width and length for three hours or so, I stopped, lethargic and light-headed. I sat under an acacia tree, leaning back on my pack, peering vacantly around. I'd sampled a few light-colored granite rocks but no dark rocks so far. The simmering haze undulated like the waves of the sea in front of me. The glare made me squint even with sunglasses on. Perhaps I had missed some rocks? It was easy to do here, as there wasn't much to differentiate one patch of sandy earth from the next. I felt a bit dizzy, which was unusual. Had last night's alcohol dehydrated me, inhibiting my body's ability to cool itself?

Hell, it's past eleven o'clock and has gotten very hot. From one of my two canteens I took my first drink since leaving the Land Rover. I tried a field trick of rotating my eyes like a strobe light in slow motion, and then focusing on whatever I looked at to see if I could recognize it. I saw scanty succulents lying flat in the sand not far away. Such things always amazed me, as all water had evaporated months ago yet the plants were vivid green. Widely spaced ones called *vygies* had grown next to rocks. Their waxy leaves crawled across the sands like witches' fingers and seemed to thrive on the sun. A gemsbok cucumber lay twisting in the sands, feeling in all directions. Frail shrubs, resembling barren twigs, were close by. No problem with basic recognition, it seemed. However, when I stood up, I was shaking.

I groped for the thermometer in my pack. *Good God, it is 105 degrees Fahrenheit!* I tapped it. Was it broken? If the human body rose to that level, heatstroke and death were approaching. Eventually you dried up and fell over, as your bodily fluids evaporated. Worse news for me was that as the sun rose higher, cruel and without remorse, it would get hotter! I'd walked too long in the heat.

The Land Rover was too far away to return to it. I would be caught in the trembling midday heat. Like an oven, it blazed down from above and radiated off the ground below, roasting everything between like meat. Hell, the sand temperature would be much higher than ambient air, probably easily in excess of 140 degrees Fahrenheit!

I told myself I would be okay. It was just a question of resting in what shade I could find and not panicking. To hell with finding rocks. I looked up; the sun was now near its zenith, high noon. All animal life stands still then or seeks shade, waiting for coolness to return to the air. Even the insects disappear and a strange silence occurs.

I was thirsty. I'd already begun to dry up. I had to pace myself.

As the sun moved overhead in the sky, any shade from the thin leaves of the umbrella trees that I might sit under would disappear—leaving me exposed, grilling like meat on the sand if I fell asleep. Closer to the tree's trunk there was a more continuous shade, but disease-bearing ticks were more likely to fall on me.

My throat thickened. Or was it my mouth? There was no saliva. The heat had completely penetrated my body. It was a dry heat too, which meant even sweating would not help cool it down. I would have to last hours out here before it got cooler. Again I tried to assure myself: *I can do it.*

Could I have heatstroke now? Oftentimes a person didn't know. You could tell if you couldn't walk in a straight line. I stood up and checked my tracks in the sand. Shit, they made a lazy "S" pattern. At least I wouldn't circle, I chuckled, as I had the river to follow. Death in a desert is often preceded by the graveyard spiral.

Getting sleepy, I wandered over to the shaded perimeter of another larger acacia tree, which had taken root in the cracks of a rock. Its stems had twisted as if in agony from the drought and heat. I sat down and stared blankly as the sun sliced through the spindly, meager leaves above. I took another drink and looked around. I didn't feel well. *Just relax, ask yourself questions, keep your mind off the heat, and don't sleep.*

Where do the plants get their water?

The desert vegetation absorbs the humidity of the cooler night, and for this reason many wild animals do their feeding in the dark. Some plants are miraculously slippery with moisture when stepped on, even in the middle of the day.

I began to feel sick to my stomach, a sure sign I was burning up. I lay back and swallowed it down, not wanting to retch and lose precious fluid. *It is one o'clock.* The sky was endless, strong and massive. It wrapped around everything, overpowered all, its strength ruthlessly magnified from the sun. I had to hold on for another three or four hours at least. I was tired and the earth was baking. My tongue seemed thicker, like a sausage. I turned my head to the side and looked forward; I couldn't see so well. In the near distance grew a weird looking quiver tree—*Aloe dichotoma*—about three feet thick and fifteen feet tall. I saw its thick, succulent, wax-coated leaves high in the air, allowing them to escape radiated heat from the sands below. *Clever plant; I wish I could escape the radiated heat.*

I continued to feel sick, just able to hold myself. It was stretching my nerves. What could take my mind off that? I rolled onto my stomach. *My forehead is cool?* Air, smelling of dust, caked in my nose and coated my tongue. I was eating it. I couldn't spit it out, my mouth was so dry. I felt my cracked lips. *Fuck, it is so barren, so dry*. The sand was too hot to touch, the rocks shivered in the heat. I needed to concentrate on something, to think.

What are quiver trees used for?

Bushmen, who have survived more than four thousand years, use the lightweight branches to make quivers for their arrows. The insides of the branches are fibrous and once removed, a strong, hollow tube remains.

Raising my head slowly, I drank several mouthfuls of calming water. *I will be fine.* Was I breathing quicker; was my heart beating faster? Things seemed out of order, as if I had lost balance. The brightness was dazzling. I looked up through sunglasses to see two suns. I blinked one of them away. My eyelids drooped, heavy. *Maybe I am worse off than I thought?*

Time is related to where the sun is in the sky. *When will it get cooler?* My lethargy seemed almost hypnotic, calling me to sleep . . .sleep . . . sleep.

I lay face down again. My head began to pound from the exertion of thinking, as though a blunt nail was being driven with hammer blows into my temple—its dull iron point pressed into my eyes. Yes, I wanted to sleep; the heat of the sand penetrated my drained body. *It is just after two o'clock.* A long way to go. I groggily looked up in another direction and noticed something about ten feet away, on the ground, level with my eyes.

I blinked repeatedly. It was thick . . . what . . . a snake! It was a fucking big snake. *Shit!* It had a repugnant flattened head, sculptured like the ace of spades—a puff adder. I jumped up and stumbled backward, falling over myself. *It must live here somewhere, probably under a river rock, probably coming home.* I stared.

My heart was pumping fast now—not good. I had a theory about snakes. You saw them, but unless they were literally on top of you or you were used to seeing them, there was a delay in recognizing them. The mind just couldn't identify a snake that fast. I figured it was because they had no recognizable limbs like humans and animals. The transmitted image didn't know where to go at first. It gave a snake the advantage of surprise every time.

He wasn't moving. Puff adders are unhurried and lazy; he wouldn't attack. It was too hot for him to be out in the heat. The sand alone was blistering, so he *had* to move. Like me, he had to move. I looked at him again as I blinked. There was a smell, a faint kind of sweetness tinged with decomposition. *Shit! He is dead?*

A snake dead from the heat? Fried on the sand? I had to get out of here. Staggering forward on unsure legs I mistakenly went up the flat-looking river instead of down toward the Land Rover. I was stumbling more, the sand felt heavy to walk through. I needed to find another tree and forget the dehydrated snake. The sun burned through my clothes. Swishing the canteen, I drank two mouthfuls. It was almost empty. Was that a tree in the haze? How did the Bushmen find food? They knew which bulbs, tubers, and other nutritious roots hidden under the sand were edible and how to find them.

I stopped to pee on a desiccated plant, leaving only a token of water. That was a good sign, as the body protectively held on to water under such conditions. *Is it a camel that pees a huge stream backward on its own long leg to help cool itself? Does he rotate his legs?* I giggled. My urine was quite yellow, which meant heatstroke. But then, everything was yellow in the glare.

I took a sip of water and pictured tiny bows and two-foot-long, frail reed arrows the Bushmen killed with. It was hard to believe. The feather-light arrow shafts are made in three segments. The arrowhead, a quarter inch wide and made from crude iron or a tin can was most important, as it was designed to enter the animal just enough to anchor itself. The rest of the shaft then fell off, so as to not allow the animal to dislodge the arrowhead itself. The arrowhead's poison did the killing. It was like catching the flu. It took a few days to work its way into a larger body system, but once it did, the animal dropped, weak and tired. *I have heat flu,* I told myself.

My exposed arms and legs were hot and dry like cooked meat. I wanted to drink salad dressing, vinegar and oil. I was burning up. My lips cracked more. Then there was the glare; it seemed to burn through my sunglasses. My head continued to pound. I sat under another tree, my head in my hands. Think about something. *The arrow . . . Maybe it would fall out?*

Once the little arrow hit, the little men—often the size of a very lean thirteen-year-old—jumped up from their hiding place and went after their victim on foot. Mile after mile they ran without rest, chasing animals as big as the horse-size eland. They could follow tracks over rock, gauge the time it took for a plant to recover from being stepped on and how long it took a spider to rebuild a disturbed cobweb. They could smell if an animal had passed. Such ability meant life or death for their small nomadic band, which shared everything.

Nothing more came out of the tipped canteen. I dropped it on the sand and drank from the other. I needed more water now than earlier. The slow movement of a tortoise caught my drifting eye. Squinting, I saw a pattern on its eight-inch-long shell. It was a geometric tortoise. I stood wobbling a bit but followed him as he plodded forward valiantly to a protruding rock that offered shade. I don't know why I followed him—I think it was just something to do, something my mind focused on.

Is the rock a metapelite? He is trying to slip under a metapelite? I focused and drank. I blinked and drank more from the remaining canteen as I stared. *It's astonishing how you find something you need when you least expect it.*

The two-foot-by-four-foot specimen lay hidden within the lighter-colored granite-type rocks enclosing it. There were large hexagonal red crystals of almandine garnet, the same as I'd seen at the far end of my field area. There was no mistaking it, even in my condition. It had geometry, like the tortoise; I weakly smiled as I looked at it, wobbling.

But yeah, it *was* a high-grade metamorphic rock. It formed at *high* temperatures and pressures. According to Afrikaner Joost, this area formed at *low* temperatures and pressures. This was a huge difference to a geologist—almost like saying an Eskimo was born in Hawaii. His dissertation was wrong. *Ha, ha, ha and here I am shrinking up. Talk about winning the battle and losing the war . . . My body will be found draped over this fucking rock.*

No, something would drag me off and suck my bones in the shade. Odd, the rare tortoise led me here . . . like the white rabbit leading Alice. Was I delirious? *It is three o'clock and still hot. When will this end?*

I took two large swallows of water. *What to do about the rock? Should I ignore it? Hell, how can I? No man believes he will die,* I remembered. *I am just thinking silly. I'm okay; just need water and rest.*

I ordered myself to photograph the rock and take a sample. I saw all the veins in my hands and arms were popping out, swollen from carrying rivers of blood in the attempt to cool it. Finding this rock was a fluke. The blazing, ruthless heat was strangling my thoughts, as dry, hot air lodged itself in my throat.

Thank God metapelite was much easier to sample than the incredibly hard metabasite rocks. All I could manage were a few hits with the sledgehammer; my head throbbed each time I swung and my elbows ached. I swallowed as best as I could to keep myself from throwing up. My body was trying to purge itself of alcohol, but I knew it was better to keep the fluid in. My hot, dry skin stretched as the rock fractured. I stuffed some pieces in my almost empty pack. The tortoise was nowhere to be seen. I sat looking at the pack. I picked it up and stumbled forward.

Did the Bushman use poison from snakes? Was it spider or lizard poison? No, it was a beetle, the larvae and pupae of a certain beetle . . .

I staggered up to an umbrella tree. My nose was blocked, so I breathed through my mouth, sucking in the talcum dust on my tongue. Probably my body temperature had increased. *It is four o'clock.* Shadows were longer, but everything was still hot, hot, hot. I was nodding again and lay down. Something stirred me. I focused slowly. It was a fat tick sucking blood on my bare leg.

Aaaaiiieee!

I hastily raised my hand to strike it and then stopped. I fumbled for a match and held it close until the tick fell off, taking its black head with it and not leaving it embedded in my skin. *I am still thinking; that's good.*

I looked up and saw a windmill well shimmering in the distance. It gave me a target. Did I expect to find water? Was it a mirage? Lurching, I reached it twenty minutes later. It was real. Unluckily, some days or weeks before, in order to save water the farmer had stopped the windmill from turning. Not knowing that, however, I leaned up and looked over its six-foot-high concrete side, squinting. *What?* Yellow fangs as long as bananas flashed at me as maniacal growls raged from the tank. I fell backward, blinking, not knowing what had happened. I was unable to see clearly. *Shit! I've seen the devil.* Was I mad? Slowly I raised my head over the rim again . . . What? . . . *A raving baboon!*

I looked again. Yeah, a mad baboon in an empty tank. He ran in circles, furious and screaming. Even if I could have leaned a log in or something to help him out, I wouldn't have, as he would have likely killed me in his rage.

He stopped for a moment and stared at me. His two fangs twisted with his head as he glared. His yellow eyes were frenzied—yeah, he was mad. Maybe he'd been so before he entered the tank? Heatstroke? He saw water in the tank that wasn't there and jumped in? Hell, first a snake and now a baboon had succumbed to this heat. What a day . . . What chance did I have?

Hell! Baboons travel together! I looked around quickly in fear. *I need to get the fuck out of here fast!*

He continued screaming as I scrambled away. My skin prickled. I was happy he was in the tank. A large drink calmed me as I hurried. Jesus, imagine a mad baboon wandering up and down the dry river, and me with only my sledgehammer. *What else can happen?* The animals were coming out of hiding now—feeding time. I wobbled back in the same direction I came. It was almost five o'clock, about a third of a canteen of water left. It would be a hard, slow walk back, but it *was* cooler. I drank more water and felt better.

A half an hour later, just as I stepped around a small granite-gneiss boulder protruding up from the riverbed, a cobra reared slowly up over a foot high in my path, its hood spread out like wings—and hissing loudly. An instant later, fear struck my mind: *SHIT! SNAKE! RUN!*

I scrambled backward, arms flailing. Hell, another snake! It was a Cape cobra, the most venomous of the African cobras. A flattened hood, bright beady eyes, its bifurcated tongue slipping in and out between lips . . . It seemed to grin at me like evil. I wasn't hallucinating, but it was probably as fucked up and surprised as I was in this heat. If it bit me, my heart would stop within a few hours. Although an aggressive snake, it normally didn't strike unless its life was in peril. I was now several yards out of its hitting range, which is about as high as it raises itself. Still watching the cobra, I stood . . . tired, bent over, gasping. I wanted to throw up again. It was too much.

A clump of Cape fig, or Hottentot fig, a succulent with edible fruit, was next to him. He was waiting, curled up in a bit of shade, to ambush any rodent that came scampering around the rock to eat the plant's fruit. Instead of a plump rock dassie, I had come over the top of the rock.

I have to get out of here. It was becoming a fucking madhouse. I shook my canteen and finished the water. It would be a long way, too long, but it was now *just past five o'clock*. I needed a field trick to help me reach the Land Rover. I broke the route up into short segments defined by trees or rocks as they came within sight. I hypothetically named the first one 100 and began a countdown to 0, which stood for the last segment containing the Land Rover. Each segment achieved was an accomplishment, a goal I marked off in my mind.

At segment 48 the Land Rover appeared; I had lost count many times. I stopped and stared at it with half-open eyes. Was it really there? It seemed days since I'd seen it. Dry mouthed and nauseated, I was sick again. This time I let it go while dumping my mostly empty pack off my shoulders, then staggered forward. Reaching the Land Rover, I leaned on it a moment to steady myself but shook trying to pull open its back door. Grasping inside, my fingers walked over the metal floor to a container of water. I pulled it to me and, shaking, tipped water into my mouth. I sat on the sand hydrating myself slowly for an hour. Warm quarts went down. I gagged as I finished one container after another. I poured it on my head and clothes. I soaked my large neckerchief again and again and draped it around my neck. So much relief . . . Like a very tight cork worked gently out of a bottle, I slowly came out of the Namaqualand heat. I was weak and nervous—unusually nervous. I couldn't drive. I knew I'd pushed it too far.

Tonight I would sleep here, on the sand. Life slowly came back to my body, and nervousness was sucked into the night. I lay down as dusk came. Waking an hour later, I gauged myself. I seemed okay but was still uneasy, with bits of nervousness and shakiness. I wasn't so sick to my stomach, however; I thought I was purged of alcohol. I ought to eat. I needed normal activities, a purpose that would make my mind work. I needed to think. A young, fit body can recuperate surprisingly quickly, but that can be deceptive too. I sat some more.

Is it over?

I looked at myself in the Land Rover's side mirror. My eyes

were clear. Yeah, I had to eat, to do something normal, get strength back into my body. From the back of the Rover I pulled my kitchen box outward and surveyed its contents. The shakiness continued, but I was in control. The day had been too long, I'd had no influence on it and had to react to it, accept it. *Now I am in control,* I said to myself again. I was part of the recovery. Recovery from what? Later I learned the temperature had reached more than *115 degrees Fahrenheit.*

There was a handful of split peas soaking in a large capped jar of water that I prepared a day or so ago. I poured them into a pot on the burner of my Coleman gas stove and lit it. To a saucepan I added chopped onion, carrot, and garlic from a container and heated it on another gas stove. I sat transfixed, and some minutes later added them together to cook more. Still weak, I drank more. The mixture bubbled and I smelled it—the base for a hearty soup. Life was coming back.

Standing, I pulled out an oblong tin can of Dutch ham. The can had a convenient opener fastened underneath that my sore fingers broke free. I emptied it onto a cutting board and cut it up as best as I could with a knife from the box. More water would be needed, as the ham was always salty. I needed salt anyway.

I grinned and chuckled, smelled the food again, and figured I wasn't hysterical, just happy and hungry.

A slow hour later, the temperature had dropped further. Rummaging around in my kitchen box, I found my pipe and tobacco. After banging the barrel to get the sand and mites out, I stuffed it with tobacco. It was as dry as paper from the Rover's internal heat. It burned up. I didn't want it anyway; I was just pretending, doing something. Sitting on a rock, I looked down, breathing slowly and steadily, it seemed to me. A large pale-yellow scorpion came out from under my seat. He turned swiftly and arched his poison-laden tail. Stumbling up, I focused and slammed my boot on it. That was normal.

The long day was over, and it was time for me to brush my teeth and sleep. *How did the Bushmen clean their teeth?* I asked, not able to stop myself. *With a fresh twig, the end chewed into a brush,* I answered. It wasn't over yet, I realized; my mind was still moving back and forth. It hadn't caught up yet. Maybe in the morning . . . I would go back to my trailer then and rest there.

Chapter 9

Unexpected Exposure: Lady and Leopard

After being in the field for several months, I needed to return to Cape Town. Loading my rocks carefully so the Land Rover wouldn't tilt and strain its heavy-duty springs, I began the slow drive back. Catina was away visiting her parents, but my new flatmate, Patrick, had returned from his field area. I was glad to see him again. When I first met him, he'd arrived at the PRU with two cardboard suitcases held together with ropes—a real Oliver Twist. At the same time he had a quick smile and bright eyes. "I was born in Donegal, Oireland," he told me. "Me father had a harse [horse] and cart and dug peat for a living." His mother had died when he was twelve. After graduating from the University of Dublin, he joined up with the Anglo-American Foundation as a way to pay off his debts. They sent him to prospect for gold in the isolated Botswana "bush" for four years.

"That's more than five hundred miles to the north from here," he instructed me. "Dr. Manfred contacted Anglo, and they sent me here to study—it was cheaper for them. And I'd become too reclusive, talking only to my dog and living with my team of Tswanas," he continued, almost shyly. "I didn't want to see anyone coming. I couldn't drive anything but a harse when I got there."

Patrick was one of those rare good guys you don't meet often. He became a lasting friend that I relied on in difficult days.

A week or so after I returned, we climbed Table Mountain. Looking down at the magnificent bay that formed a shimmering water necklace leading to the city, I realized just how lovely this area of the world was. On windless days, when cumulous clouds were low in the sky, they covered the flat mountain like a tablecloth, pillowing off its sides, and we walked in the sky. Nearby, high hills formed a resting lion with its head raised: a dramatic entrance to Africa from the sea.

Early sailors, including Charles Darwin, called Cape Town the "Tavern of the Seas," as it was a great "inn" on the way to the east. Bartolomeu Dias, the famous Portuguese explorer, first recorded its existence in the mid- to late 1400s. Following that, tall-mast ships came into the bay, their sails billowing while tar-haired, diminutive Portuguese sailors climbed high in the riggings. A colony began in the mid-1600s with Dutch-born Jan van Riebeeck serving as its administrator, at which time it became primarily a harbor for two- and three-mast Dutch fluyts, which were cargo ships sailing to India and the Far East. By bringing in slaves from Jakarta and Madagascar on those ships, van Riebeeck began building a rainbow community, with the establishment of a full settlement at Cape Town. Concurrently, he established a class of permanent white settlers who lived apart from others. That was the beginning of apartheid.

Talk then turned to our former PRU director. "So Dr. Manfred returned to Germany," Patrick said to me as we walked down the mountain.

"Yes. With all his get-up-and-go, he found and brought us here, started us off, promptly abandoned us, and is now head of a geology department in his home country."

"What do you think of Joost, the new head of the PRU?" Our new boss was none other than Afrikaner Joost.

"I'm worried. It could turn out to be an uncomfortable experience for me. He worked in my field area," I replied in a somber voice.

"But that's good news. He should be happy you are working in his old area!"

"It certainly should be that way, but so far I can't find any evidence that supports his PhD dissertation concerning what happened geologically in that area. I fear I will be at odds with him. At this point, all I can see is that what he says is incorrect, Patrick—dead wide of the mark. But I will be assigned a new advisor now that Dr. Manfred is gone. He should help with this."

Patrick looked at me as if he didn't believe my blunt statements. I couldn't blame him. I didn't want to believe it myself. It wasn't supposed to happen this way.

I continued. "I did indirectly broach the subject once with him and was told bluntly that he was correct. '*I'm sure of it*,' he stated, staring into my eyes with conviction." Shit, he was a professor. Still, this was a university and truth, supportable truth, should conquer all. Yeah, I needed an advisor.

"Let's hope you're wrong," Patrick said without conviction. He was picking at his black beard as if there were lice in it. "Maybe avoid any confrontation for as long as possible until you're very sure."

"Yes. He is an Afrikaner. They don't turn easily. Still, maybe he will be gracious, and maybe I am wrong." I shrugged. I didn't want to make problems.

We continued walking down a steep portion of the mountain. "Come on, lad, tonight we'll go to a nightclub and have a wee dram."

"It would be good to get my mind off this."

Portuguese Catina and I had not seen each other since I'd been in the field, and she was now doing fieldwork herself. Patrick and I went to a local drinking place where I met twenty-year-old Margret, a long-legged, slim blonde of British descent. She was a sophomore at UCT and wore a simple, short, white-satin-sleeved dress that highlighted her nice figure. Dancing until late, she took my mind off my problems. It was a much-needed relief, so when she invited me to a party her part-time employer was hosting in a few days, I accepted. Her employer, it turned out, lived in a ranch-style home with windows and doors that opened to a swimming pool tucked away in a secluded portion of the backyard. The African summer night was warm, and large trees swayed in the quiet breeze. Lights from the house barely reached the pool's still surface.

"Come, let's swim," Margret said, pulling me away from her heavily jeweled male boss moments after we arrived and before anyone could talk to me.

"I have no swimming suit."

"Neither do I," she said, as if we were discussing handkerchiefs. She led me to a darkened area adjacent to a huge tree. "Would you mind unzipping my dress?"

"No problem," I replied, thinking of the sight . . . *I'm like a falcon getting its hood slipped off.*

We slipped into the water, coyly paddling around in decreasing circles. Most of me remained a gentleman even as we converged and embraced. After emerging from the pool, I toweled her dry in the shadows.

She watched me as I lingeringly patted here and there, and she curled her toes as a bird would on a wire. Her lips parted but she said nothing. My mind raced, as did the blood in my body. I lost control.

She was young and looked as innocent as a fawn. I suggested we go to my apartment. She nodded gently yes.

What happened next was like peeling a banana; once you start, what is the point of stopping? "You will be my third man," she said. Odd information, I thought. "The other two were of no consequence," she added.

"No problem."

I saw Margret two or three times after that over the next two weeks. We walked and talked as young people do. In the evenings, after I'd left the university, I'd pick her up at eleven o'clock or so from her job in central Cape Town. She told me she worked in real estate, and late-evening appointments with clients were the norm. Sometimes she came out of her building red-faced, as though she had just taken a sauna. I eventually figured out she actually had—that she worked in a massage parlor. The real estate story was untrue, "a little white lie," she called it.

Dear God, what a jolt. It left me confused for days. I didn't know about such things; I'd never been inside that kind of place and had enough problems of my own to deal with. I missed Catina and decided not to see Margret again, though I had to admit she sure as hell got my mind off Afrikaner Joost for a bit. I began to wonder things like if I'd be punished from above for that, and if Joost's dissertation would be considered a little mistake also.

Back at the university, I retreated to my office to sulk in confusion. I decided to continue keeping quiet about my contradictory geology findings. Margret made me realize I was naive. What did I know? Maybe I was wrong.

Anyway, it was soon time to return to the field. There were large areas I hadn't yet sampled, like the stretch I called the Canyon. I'd driven through it several times on the way to the coast and elsewhere. It had been carved out by a river that predated the currently dry Buffels River that now bisected it. It was an interesting enigma in the dominating flat, open lands.

A week or so after returning to my field area, I arrived at the Canyon early one morning. Its jutting, steep, irregular walls were made of light-colored granite-gneiss Precambrian rock. Were there darker metabasite rocks or metapelites here? If there were, according to Afrikaner Joost they would be in one of his zones where, during their formation, the temperature and pressure were *lower* than farther inland—or anywhere else, for that matter.

The cool blackness hung around me as the sun first peered over the lofty canyon rim. Like a scythe of golden light, it swung forward, chasing the retreating dark enemy. Erosion-sculpted rocks changed into brilliant yellow, gold, and then orange-brown as the night scurried away, helpless and insubstantial as a cloud. I felt a sense of awakening.

The flat-topped and thorn-laden acacia trees were broken into twisted, gnarled, and stunted symbols of defiance and eccentric beauty by ravaging canyon winds that had blasted them endlessly with storms of sand. Like tattered green umbrellas, they offered only spotty shelter from the relentless sun.

Closer to the ground was gray-brown straggling, stunted mimosa scrub with tiny featherlike leaves and robust thorns, inches long with fire-hardened spikes, growing at whatever intervals they could suck enough nourishment from the shriveled, parched, sandy soil.

I parked my Land Rover near the narrow canyon entrance and began walking down the indistinct, snaking sand track directly adjacent to the dry river. Sometimes it wound across the river. It was not a trail intentionally constructed but rather the product of larger animals, like cats, prowling in and around boulders and thorn trees as they foraged for food on their way through the canyon.

Several hours after dawn, a second transformation occurred. Shimmering veils of heat steamed upward from the warm sand. With each degree increase in temperature, the extraordinary landscape turned into a mystical world of ever-dancing shapes. It seemed that lofty music should accompany this transformation. Instead, an immense calmness prevailed. A soft breeze ruffled the parched scrubs. It was all like magic and left me marveling how ingenious nature could be as I walked around boulders in the sandy riverbed, searching for rocks.

During the annual torrential rain, when the calm African sky was driven mad with raven-black storm clouds, the river filled and flowed, sweeping away portions of the path along with any uneaten bones and flesh that scavenging jackals and hyenas left behind. The night after the purging water subsided, the animals would make a new path in virtually the same favored and logical place. When humans came to the canyon, they simply followed the path and eventually broadened it into a thin ribbon of dirt, sand, and rock.

The larger animals were now gone, preferring more remote places to those touched by travelers and farmers. At least I *thought* the large predators were gone.

Shifting my backpack for balance, I climbed up and down boulder-laden slopes, hammering fist-size samples of rock off the most interesting outcrops. I couldn't find metabasites or metapelites but sampled most every rock type I mapped as I went along. It was tricky when the slopes were especially precipitous.

As usual I numbered each sample and recorded its location on an aerial photograph with a pinprick. On the reverse (blank side) of the photograph, I labeled the tiny hole with the specimen's number.

Removing the increasingly heavy pack was like dropping an anchor from a boat. To put it on, I had to sit on the ground and work my arms into the straps, bend forward, and then struggle to stand. It would be easy to lose my balance and break my leg.

To counter the increasing heat, I wet my bandana handkerchief with canteen water every hour or so and wrapped it around my neck. I worked steadily all day, stopping only to rest and eat dried strips of biltong, fruit, and peanuts.

Then I saw a metabasite rock, a fine specimen about the size of a small bathtub, engulfed by white granite-gneiss rock, kind of like a head surrounded by a soft pillow. I hit it with my sledgehammer. It rang like a bell.

Oh dear God, I thought. Another unexpected hard-ringing *high* temperature and pressure—from black rock that's not supposed to be here, but here it is. I looked at it closely. It *was* more evidence contradicting Afrikaner Joost's findings. I took off my nearly dry bandana and wiped my face. Well, here it was, and I had to deal with it. I broke off a large piece.

With less than half an hour before darkness, I decided to call it a day. Picking my way down the canyon slope, I left my burden next to the sand road. I planned to walk the half mile or so back along the riverbed to where I'd left my Land Rover and drive back later for the specimens. I carried my sledgehammer as usual, in case I decided to take another sample. The moon was full; it would illuminate the course of the flat sand-colored river when the sun set. Stretching from the freedom of the pack, I looked forward to a nice walk in the moonlight.

That was not to happen.

It began with a cough—a guttural, rasping yowl. I stopped. The odd sound stood out from the normal quiet of the evening. Minutes passed. An impatient and dramatic expulsion of air punctuated the calmness. What the hell? Moving slightly, I scanned each boulder and ledge as best as I could. No longer lulled by the pleasantness of the valley, I sensed change.

Suddenly, some two hundred feet away, I saw eyes: a pair of yellow-jeweled, patient eyes staring at me . . . the eyes of a creature used to the solitude of Africa . . . a leopard. I froze. The safety of my Land Rover was still at least half a mile away, and I would be following a twisted path. Not good . . . Then again, perhaps the leopard wasn't interested in me?

Don't be naive.

Partially hidden on a ledge in the direction I was walking, the leopard stood immobile like a solider, watching me. Damn! Where had he come from? A den between boulders surrounded by skeletal brushes? A nocturnal and solitary animal, he had probably descended from the heights in search of food.

Was I somehow in his territory?

Hadn't all the cats wandered to the far north long before, as farmers and hunters moved in and chased them away? Something had made him come here. Perhaps he was lame or too old to hunt wild game and was now after sheep and goats or whatever weak prey he could find. Cats get addicted to the taste of such animals. Shit, that category could soon include me, although normally they didn't attack humans—did they? Frightened, I tightened my hold on the sledgehammer riding on my shoulder.

Abruptly his tawny shape vanished. As I blinked, he disappeared into the boulder labyrinth like a shadow before the sun. All was silent except for the beating of my heart.

I looked around. No farmhouses were anywhere near. I had no choice but to continue walking. I fought the immense desire to run. That would probably only entice him. *Never run*, I told myself.

Some fifteen scary minutes passed. He could be behind every boulder. A rock rolled as a paw jarred it. He reappeared a hundred and fifty feet away. His back radiated sunset gold—a magnificently made animal capable of ripping me to shreds. I couldn't take my eyes off him. In another moment, he disappeared.

It became profoundly silent around me. The silence of a grave charged with menacing suspense. Even the cicadas had stopped chirping. The cat *was* hunting.

Now the canyon looked different, like a great jagged rip in the plateau, a breached portion of the earth's body, exposing a shriveled vein of a river surrounded by organ-like boulders. All reflected the red of the setting sun.

The air was cooler now, but I felt drops of sweat on my back. The sun touched the canyon rim, and it looked like a molten tangerine cannonball about to be dropped in black ink. Long finger-like shadows grew around me. Then, in an instant, light and weird shadows disappeared. The African sun departed like that, especially from inside a canyon. As expected, in a moment the white-green light of the high moon appeared and cast odd phantom shadows in the riverbed. It would take about twenty minutes for my full night vision to settle in, but I sure couldn't sit and wait for it. The cat was following me and would not depart; hungry or curious, he would come closer and closer. There was no place for me to hide; this was his canyon now.

I looked up and saw clouds, clouds that might black out the frugal light I depended on. The dark of an obscured moon was favored by leopards; a leopard's moon, some called it.

The wispy sounds of the dying evening had indeed disappeared, except for those made by a southern evening wind as it tortured the dehydrated acacia tree leaves, making them twist and turn. It had become a nightmare place.

I continued to walk fast but not so fast as to show panic. A cloud passed in front of the moon and the river sands went shadowy black. The emptiness became eerie. My eyes tried to adjust, taking in just a bit more light. I could vaguely see silhouettes of spindly trees and bushes and kept up my controlled pace. Would he smell my sweat and sense my fear? Would he prefer the blackness?

I couldn't stop thinking about him ripping at me. Who could? Before a lion attacks, he gives warning. A leopard is not so cordial. He just attacks.

Suddenly I felt the drag of a thorny branch across my jeans. In the silence, it sounded like a branch ripping off a tree. My heart leaped. I jumped back and jerked up the hammer, ready to swing. My body was flooded with adrenalin; it was as if I couldn't breathe.

Nothing, no sound. I took a ragged breath, and gradually my heart rate slowed. I realized that I had strayed too close to the needle-sharp mimosa scrub, diabolically designed to dig into flesh. It had torn my arm, and blood oozed down. As I looked at it I heard, much closer now, a gruff *haugh-haugh* cough. Had he smelled my blood and sensed my fear? I kept walking.

The clouds shifted partially away from the moon, and I was able to see the riverbed more clearly. I searched for tawny things behind each scrub . . . and I saw his pugmarks where he had passed. I was really sweating; I felt perspiration running down my back between my shoulder blades.

Fewer than a hundred and fifty feet in front of me the perfectly designed carnivore slowly crossed the riverbed where there were more boulders. The DNA in his genes was ten thousand years old—he was driven by an unforgiving force. He walked like a conqueror. His body was so sleek, I wondered if he was actually thin, hungry. I held my breath. His breathing was loud, alien, rasping, and powerful.

The vertical pupils in his glowing, unblinking eyes locked with mine for a brief moment as he weighed his prey. Did his tail twitch? I visualized myself swinging the hammer—with little effect except to make him more ferocious. His unshakable eyes would blaze like hot coals while his filthy fangs and scimitar claws raked me. I would scream. What was left, bleached bones and ragged clothing caught on thorn trees, would be swept away when the river rose again.

Just then the moon broke completely free of the clouds, illuminating the river sands. The leopard did not move. Leopards are cautious animals. Was there too much moonlight? Did its eerie brightness in the dark confuse him?

He veered off toward a cascade of boulders and leaped up them a half-dozen feet at a time, like an uncoiled spring. Resolve was carved into every piston thrust of his powerful legs. He didn't look old or lame then. Hell no . . .

The sharp acid taste of fear hung in my mouth. A soft breeze rustled the bone-dry, spidery undergrowth. I breathed deeply, trying to calm myself, but my heart was pumping. He remained between me and my vehicle. Did he sense I was going there?

I looked around again, no rock cavity to crawl into to protect myself. I had to reach the Land Rover before something happened.

My mind was fixed on the Rover: Once I got inside, he couldn't hurt me. I forced myself to walk steadily and not too fast. Everything I did now was essential to survival. I saw the silhouetted canyon entrance some three hundred feet away.

Closer . . . I was near where the cat had jumped up the boulders—the last place I saw him. I had to be brave. The thin dirt track snaked ever nearer to the riverbed. The wind strengthened as it was funneled through the canyon's mouth.

This was the moment.

Would he pounce?

Focusing on the white metal patch in the near distance, I realized the canyon was getting darker . . . another cloud! Step by step I walked—my Via Dolorosa. A hundred feet to go.

Then I heard the cough.

I lost control and sprinted toward the vehicle.

Did I lock its doors? Oh God!

I darted forward, knowing I was no match for the cat. Leaping over small boulders and skirting large ones, I ran. Only when I was very close, a few feet away, did I hear a sound behind me, the shifting of a loose rock as he struck it. Was it him? Was it my imagination? Was it my fear?

Shit! I whirled the hammer behind me without looking and slammed into the Rover, grabbing the door handle, feeling that at any second he would sink his claws into me and drag me backward. My heart was exploding.

I yanked the door open, dived in, grasped at the gears, anything to pull myself forward and hold on to while I kicked my feet viciously, thinking he was behind me. My eyes were wide and unfocused as I yanked myself farther inside and jerked my legs in after me. At the same instant I twisted around and yanked at the door.

"Shit! Shit! Shit!" I yelled as I slumped over the seat and steering wheel, gasping. My heart was beating like someone hitting a drum.

Finally, I stared out the front, then the side windows, trying to calm down. The thump of a pouncing cat at the Rover hadn't occurred. I was safe inside; it was as if some frantic panic had suddenly been released. Perhaps the leopard had only been playing a game of cat and mouse. If he had wanted me, he could have taken me right then. He wouldn't have let me get away.

I found my key but it took several tries before I could slide it into the ignition slot. Then I pulled the headlight switch on to high beams.

The lights stared down the canyon. The cat was gone. I lowered the window an inch or two and listened. Nothing.

I drove back to my trailer, delighted to sleep inside that night. In the morning I returned for my pack. It stank of urine.

Chapter 10

It Could Be Worse

Certain things in life are what they are; you just have to deal with them. That doesn't mean you can accept or forget them, however. Sometimes they stay with you forever.

Sitting next to my campfire one thoughtful night, I realized I'd become increasingly anxious about the looming confrontation with Afrikaner Joost. I had to face the reality that it would happen, and I needed to prepare myself. For example, it was important to be as lucid as I could and that I presented my facts logically. Were there things I didn't know? I slowly reviewed the geology of my field area step-by-step.

The billion-year-old rocks I studied originated deep below the surface of the earth, somewhere in the neighborhood of eighteen miles. Over those billion years, the rocks eventually rose to the surface by the natural processes of uplift in conjunction with erosion, causing the removal of the existing overhead surface rocks. Due to their depth within the earth, where movement and deformation always occurred, and powered by heat from the earth's core, I'd calculated from my mineral chemistry that these rocks had been subjected to uniform temperatures (T) of about 830 degrees Celsius (C) and pressures (P) of about 8 kilobars (kb).

According to Afrikaner Joost, the T and P conditions under which the rocks in my fifty-mile strip of land initially formed *changed gradually, forming zones*. At the far inland end was the higher T and P zone, and at the coastal end were far lower conditions—a difference of some 700 degrees C from end to end.

Joost also believed they were formed during only *one deformational event*. By taking the strike and dip along forty-five miles of the fifty-mile strip, I'd found it to be consistent from east to west, with a 90-degree C dip—meaning only one deformation had occurred. However, over the last five miles in the coastal region, the strike, over the distance of a dozen feet, changed to north-south, which meant a *second*, but much later, deformational event had overprinted the east-west strike; and the associated rocks had all been weathered into crumbling condition.

In geology there is an established process that explains findings as Joost postulated. But I found no supporting evidence that this process had occurred within the area I studied—and there was simply no way I could say it did. I went over my story again, thinking about the simplest way I could demonstrate my findings so no one could argue that my calculations or the things I'd done were wrong.

The bottom line was simple. I had thin sections from the metabasite rocks I'd found along the length of my field area. As I viewed them sequentially under a microscope from east to west, I found no change in the minerals present. And, in the field, as I traveled west along my field area toward the ocean, I saw no visible changes in the rock types. The black (high-grade metamorphic) metabasite rocks, oblong like partially flattened wagon wheels, were so hard and dense they rang like a bell when struck with my sledgehammer. These metabasites more or less marched along the strip, their strike east-west and dip vertical, without change. I found them like black stains across a white sand tablecloth.

The exception was near the coast, where the strike changed very abruptly to north-south, and hitting them with my sledgehammer produced a dull thud as the decomposed rock absorbed the blow.

The metapelites changed, too. New minerals formed; there was no more clearly visible red—the high T and P garnet. Over the coming days I began to think each bell-ringing hard rock was my dissertation's death knell. Looking for something that *isn't* there takes far more time than looking for something that *is* there.

On the one hand, I'd done my job and discovered something new and previously unknown. But on the other hand, I'd unearthed enough rope to hang myself. Yeah, there was no way I could bend my findings to agree with his. The differences *were* like day and night. Yet so far, he didn't seem worried—in fact, he didn't seem to care. I was nothing more than a pesky fly.

Yet I wondered: *Did he expect me to falsify reality? Maybe.* He was, after all, Afrikaner Joost. Later it reluctantly occurred to me the man simply *didn't* understand metamorphic petrology. He didn't question what was really there; instead, *he put forward what he wanted to be there.*

I returned to UCT, where I pored over everything again. I looked at thin sections along the length of my area again and again. No difference from metabasite to metabasite; the same indicative minerals were present. I confirmed it for myself: I was right, and there was no way a scientist could conclude differently. Joost's dissertation should have been rejected. I saw him as one of those old-school Afrikaners who dragged misguided opinions with him. He knew he was right, because God told him so. But what would the university decide? Surely right was right—and I wouldn't be sacrificed to protect him, would I?

I felt I was walking a thinning plank, carrying a heavy metabasite rock with me. Below me the swirling cold, deep water of the Cape Bay was calling. My major responsibility was to construct my dissertation—difficult enough to accomplish—not battle with this travesty of science. I was carrying unfair weight and staggering from the load.

Thanksgiving time in November, I realized one of the worse things about this was I couldn't, at that stage, talk to most of my colleagues about the problem, although they had all heard something about it from Patrick and Catina, who tried to defend me. They too were working on degrees and didn't want to be involved with anything that could by association adversely affect their careers. I found myself effectively removed from society by matters of diplomacy and survival.

I walked alone most days, thinking about my situation. At one depressing point I realized that if I did die from that leopard or from thirst, Joost would win. Oddly, a few days after that low point of consideration, I was invited to attend an autopsy. Given my depressing situation, it kind of fit in. Even more oddly, it helped me put things into perspective later.

* * *

The story began early one morning under a camphor tree on the beautiful Stanley Street brick sidewalk in Middle Campus, when I met a woman wearing a short summer dress, almost like a figure-skating outfit. She was walking her Scottish terrier, Arch, who seemed to enjoy pooping on whatever he could and proudly sniffing his coiled accomplishments. I took the opportunity to introduce myself.

Janet taught at the university and was an attorney of British descent. She was about ten years older than I, slim, intelligent, wore pleasing makeup, and had lovely coiffed black hair. I told her a little about my work, and found it refreshing to have someone new to talk to. I liked the clever way she spoke and suspected she was interested in me a bit.

After a week or so she casually invited me to accompany her to watch an autopsy, as if asking if I would like to have tea. "My law students have to be knowledgeable about autopsy procedures in case there is a question in their future about a murder. You can come along with us, if you like."

Of course I knew the purpose of an autopsy, but I never thought about what actually took place. I think most people don't have a reason to consider the details. However, I guessed it would take my mind off my own problems, so I accepted the invitation.

In a few days I met Janet and about ten of her students next to a shabby loading platform in the rear of a local nonwhite hospital. That loading-platform beginning should have given me pause.

"You mustn't be shocked when you see the body," Janet suddenly said. "It will be dead, remember."

"I hope so," I said. "Will this be gruesome?"

"To meet the objectives," she said, "certain things must be done. It's like preparing a turkey."

"A turkey?" I looked at her with great surprise. *An odd statement* . . .

She turned a bit red at her analogy, but then I understood she probably had the upcoming American holiday on her mind.

No further comment was made as we passed through double doors and along a dismal concrete-block corridor to the operating theater. Meeting us at the stage door was a white forensic pathologist wearing a white lab coat. He would perform the examination and officiate. He was an older man and seemed pleasant enough when Janet introduced us. We all followed him inside.

My eyes immediately fixed on the naked body of a young African girl, about twenty, lying as if asleep on a slightly slanted galvanized-metal table. A thick block of well-worn wood supported her neck. Droplets of water remained on her, as she had just been hosed down. Under different conditions it could have been heavy sweat. Her long hair hung over the end of the table, and I couldn't help thinking she was pretty. Any minute she might sit up and talk to us, it seemed. It was very unnerving.

"The air conditioning's stuffed," the doctor explained as twelve of us found places on the bleacher-type seats in the very small, hot room. "They promised to fix it, but we're obviously still waiting." His apology sounded almost as if we were entering his home.

Janet herded me to the front row and we sat down on hard seats. The body was two arms' lengths away.

"We couldn't get much closer," I whispered.

"I have to lead the way, as students don't like to get too close. If they sit way in the back, they may miss something. And it's necessary they understand the entire procedure." She placed her finely manicured hand with dangling gold bracelets on my knee—to comfort me, I assume. Actually, it felt oddly romantic.

"My assistant will now open the cranium," the pathologist stated, nodding to his Indian colleague.

The assistant decidedly leered. He knew something, I suspected. A moment before, he had looked appropriately morbid. Now he was grinning. All this seemed strangely casual. I stared at the girl and wondered how she could have died.

"I hope this isn't going to be macabre," I whispered, though I realized it couldn't be anything else.

"She is so pretty," Janet replied. "Her breasts seem full."

"Full?"

The assistant, with a stroke of his scalpel, deftly cut a curved path around the back of the girl's head. *It's peculiar she didn't scream,* I thought. At least it proved she was dead. Without delicacy or the slightest effort toward tenderness, the assistant pulled and then peeled the skin and hair banana-style over the top of her head and down to cover her face.

Dear God! I paled and felt the unease in the room. All was deathly silent. It was too much.

The assistant grinned again—a gesture I was now very suspicious about—and with extraordinary showmanship, he flicked a switch, and an electric buzz saw whined. The girl lay there. I feared some vicious stuff.

He touched the blade to the skull, which produced a hideous sound. Seconds later I heard a deep, guttural *urrrrrr-urrrrrrr* behind me. Something splattered on my neck. Startled, I half turned to see a female student jerking out of control, being sick, her hands cupped under her mouth, bent over and going for the door. She was the first.

I swallowed repeatedly to try and keep the contents of my stomach where they were when I arrived.

The saw changed pitch as it caught hold, and all attention was riveted forward . . . most through parted fingers clamped over eyes and mouths trying to hold on while others made for the door, bent over, chucking.

Janet began to knead my thigh.

Suddenly I understood the purpose of the block of wood; it would prevent the blade from striking the metal table. The sound stopped as half the skull parted from the remainder like a hard-boiled egg struck midway with a knife. The half-skull fell off the small rectangular metal table to the floor, to rock on its round side. Instinctively, I jerked my leg up to avoid contact with it and in doing so pressed my knee against Janet's. She rubbed back.

The doctor said something about coloration while pointing to the yolk-like brain that protruded from the remaining skull. While this bit of knowledge was being digested, the assistant briskly slit his way from her groin to her throat, using the buzz saw to deal with the breastbone area.

The doctor used his hand to fan the air above the chest cavity up toward his nose. "Most poisons we deal with leave a detectable odor," he said as he fanned. "I find nothing suspicious here. She died of natural causes that may or may not have been related to this." With the instincts of a magician, he extracted a fetus by its foot—without further comment.

The room was dead quiet. Mouths hung open. People even stopped gagging as they stared at the unborn baby. I turned sideways, away from the sight. One or two female students were shaking. Hysteria was near. Janet was breathing hard, like a train, her lips parted, her teeth showing with tension. It was horrifying. Our legs were pressed and rubbing—an impractical joining together rather at odds with the gory dismantling before us but, surprisingly, stimulating nonetheless. *My God*, I thought, *how low I've gone.*

"How old is the fetus?" someone managed to say, more out of shock, it seemed, than concern.

"Just a second," the pathologist said. He gripped a leg as if it were a piece of rope and lopped off the foot just above the ankle. "Seven months," he replied, peering at the stump. As he looked up, he unceremoniously released the fetus, which settled in a heap of odd-angled limbs. By this time, half the students had stumbled out of the room in bent-over positions. I was barely hanging on.

Surprisingly, someone asked another question. Actually, I reminded myself, that was what they were supposed to do. I myself could think of nothing but the immediate presentation. Apparently, the answer required a diagram. Although the subject's body fluid had been previously drained (the table sloped toward a hole), some remained and had collected at the deep end. The doctor dipped his finger in this and, with a grin I was becoming wary of, wrote on the white-porcelain tiled wall, returning his finger to the makeshift red inkwell as required.

"Could the fetus have been saved?" a male voice asked, perhaps disturbed by the outcome.

For once the doctor didn't smile. He thought for a bit and then said, "*Probably* not."

I ran my eyes over the contents on the table. The class ended, and all the organs were informally piled back into the body cavity, including the whole brain, which had been scooped out like ice cream during the proceedings. The severed half of the skull was retrieved once the doctor asked for it to be replaced. The skin was drawn back over the face like a glove put on a hand. Had the corpse been dressed and some perfume added, anyone viewing the body would not know her brain was in her stomach area.

Janet and I filed out into the warm sun and fresh air with the remaining students. There were a few nods, but no conversation. Everyone walked off quickly, trying, presumably, to put the experience behind them.

As Janet and I moved away, I wondered about the unborn baby. I didn't even know if it was a male or female. It was nothing to anyone, I realized.

The doctor said it *probably* could not have been saved. The girl was black, and the system was apartheid. I mentioned to Janet that the operation seemed a bit crude.

"The main objective for the law students," she said, "was to understand that once you are dead, you are dead—and to find out the cause in whatever manner, so you can proceed to help the living." She took my hand. "You have to carve it to make use of it, like a turkey."

I figured she must use that enlightening description with her students to open their eyes to the finality of death. I couldn't think about anything for hours after what we'd seen. That poor young girl and a baby that would never breathe.

My life could be far worse. Why worry about a bunch of rocks?

Chapter 11

Watering Hole Death

Patrick and I had a beer on the Springbok Hotel porch to celebrate the beginning of our adventure. We had talked about a trip like this some time ago—a visit to his northern field area. We would meet in Springbok and drive his Land Rover on N-7 over dry, barren land toward the Orange River, which forms the geographic boundary between South Africa and his area of study in South West Africa. He explained to me his findings. I asked him how he was coming along with his degree.

"I'm basically working on my own," he said. "No one is really advising me."

"Joost should be doing that."

"Yeah, but he's not, at least not yet. I'm collecting information and mapping my area for the moment."

"You need to talk to him, find out if you're going in the right direction. Surely he will help you."

"I guess, but I want to study my area a bit more before I present something. How are you?" It seemed as though he wanted to change the subject.

I told him, wondering if he was also having difficulties.

"I'm being viewed as if my findings were not only wrong," I said, "but that I lack the ability to investigate my field area. No one in the PRU has said the word 'dumb' (I do wonder sometimes), but it hung in the halls like a cloud pissing down on me as I passed. Some looked at me with pity, likely glad it was me and not them. I felt dejected—terrible, actually—and wished it were all my imagination. But in truth I was in deep shit, period. I need to clear my head."

"Well, this trip should give you something else to think about."

"I hope it makes things clearer, at least as far as what I will do next. I don't want to make a decision I would regret. I tried to talk to Joost, but with a smiling mask-face, he continues to state he's right. He has denied me access to the analytical machines I need to use so I can continue my calculations and comparisons, because I need more verification. He implied they were needed more by others."

An awkward silence followed. Then Patrick asked, "Could he be *that* wrong?"

"That's a problem; it's too hard to believe he's *that* wrong."

"Does he agree with anything you've discovered?"

"No. He acts convinced beyond doubt that he's correct. As you know, the new head of the geology department was automatically appointed to be my advisor when Dr. Manfred left. The new guy patronizes Joost, because Joost brings in money to the department. I'm just getting to know him, but he doesn't encourage discussion on subjects unfavorable to Joost. Hell, he was a scientist at NASA—and now he's trying to fit in here. Joost has his ear. How can I argue with someone from NASA? I need more proof of what I'm talking about. Actually, I think Joost saw an outcome he wanted and then manipulated his way to that point. Exactly what you *don't* do in science."

"Money talks everywhere. It sounds like Professor NASA has personal ambitions that would be harmed if he got on the wrong side of Afrikaner Joost. It's nothing to do with facts such as yours."

"Yeah, it sure looks that way," I said. "But hell, this is a university. That shouldn't happen here."

Patrick smiled. "The optimism of our youth."

We finished our beer, climbed into Patrick's Land Rover, and headed off toward the Orange River.

The sky was cloudless and watery blue. The land of sand and gravel lay prostrate in the unforgiving heat. Hours later, in the far distance, a mirage-like green line appeared across the horizon. As we neared, the apparition revealed itself to be the graceful and delicate tracery of fine-leaved fever trees rooted on the river's banks. Always growing near water, fever trees were blamed by early settlers for causing malaria, while their bark was used as a preventative against the disease.

We drove parallel to the trees and looked through their branches as they dipped low toward the brown water, festooned with multitudes of swaying weaver bird nests, like bulbs on a Christmas tree. The black-faced and very social birds screeched endlessly as they hung upside down from the basket-shaped nest above their heads. There on the other side of the trees we clearly saw the mighty, muddy Orange River. It looked quiet, content, as though it was hiding—a powerful snake shifting forward as it invaded and swallowed the dry land in its path.

The Orange is the longest river in South Africa. Sitting in the Land Rover, I followed its course with my finger on our map. It originated in the east, in the rugged nine-thousand-foot-high mountains of the kingdom of Lesotho. From there it ran west to the arid regions of the southern Kalahari and Namaqualand, through large areas characterized by diamond pipes. Then, on its last stretch, it reached the Atlantic Ocean, having defined the six-hundred-mile South West African border.

The gravel beds of the last fifty-mile stretch were comparatively rich with diamonds picked up and rolled forward by the ever-moving river. At the mouth was an alluvial (relating to unconsolidated sediments) diamond mine. Who would guess diamonds, dropped there by the river, were tucked away among bottom stones and mud?

Staring at the river, I remembered my boyhood cabin where I sat on the porch, dreaming of rivers because they often formed boundaries. Crossing a boundary meant you were going somewhere.

"Why is it called the Orange River?" I asked Patrick as we got out at Vioolsdrif, a very small border *dorpie* (town).

"Here at Vioolsdrif," Patrick replied, "the river goes through a narrow valley lined by high walls of rock. Temperatures reach over 100 degrees Fahrenheit. The rocks appear orange, brown, and other colors, depending on the sun. But it was named in honor of William of Orange by one of the military people who first passed through this way."

"What about Vioolsdrif? That doesn't sound British."

"Someone with the name Viool lived here. He guided travelers through the river at the shallowest places, which, in Afrikaans, are called *drifs*. Therefore it became Vioolsdrif. Look, there's the bridge across the river." He pointed out the window. "It was built maybe twenty years ago. Before that, people swam or rode horses across with Mr. Viool."

It seemed a simple place; no statues or squares to mark an occurrence. I said as much to Patrick.

"Yeah, it's the 'bush.' It has a thousand million years of undisturbed plant and animal growth, which is probably of more value than anything man could leave."

We pulled up to the customs office on the South African side and were passed through with minimal concern. On the other side of the bridge was the settlement of Noordoewer, with rows of old green-leafed grapevines.

I was amazed. "Wine—out here." The border guards were waiting for us and curious. "We are students, collecting rocks."

The guards looked at the inside of our Rover filed with camping gear. "Pull it all out," they said, looking for weapons. They warned us about going too far north, where the anti-apartheid South West Africa People's Organization (SWAPO) rebel army was active.

"It's not safe," one of the guards told us. "SWAPO is ruthless."

"From time to time they come south on raids," the other added, "and if you are in their path, you will be killed."

"Oh shit, Patrick," I said quietly as we loaded our gear back in under the guards' watchful eyes. "I didn't realize the rebels were raging this far south."

"They have no borders. Like disease."

Across the Orange River, the Namaqualand I knew became Great Namaqualand, the far western part of the Kalahari Desert. Parallel to the western coast was a seventy- to one-hundred-mile-wide belt called the Namib Desert. It was one of the driest deserts in the world, with red and orange sand dunes approaching one thousand feet in height, among the tallest in the world. That was where we were going.

Patrick reminded me the Namib is a desert because the Benguela Current, which runs north from Antarctica along the coast, captures and condenses humid air and thereby prevents it from being blown inland. It also carries diamonds discharged from the Orange River at its mouth and deposits them in the forbidding Namib. We hoped to find a few. The desert's southern border is formed by the north bank of the Orange River. Its huge, windblown sand dunes can't march across the river, so the sand grains are washed to sea. There is very little vegetation, but when rain does fall, grass seeds that have lain dormant for years briefly come alive.

We carried on, driving along the poorly defined national highway, named the C13 highway, which led north-northwest parallel to the coastal Namib Desert. It was not well traveled in the 1970s for a number of reasons, including the rebels we were warned of plus migrating sand. Then we reached the spectacular Fish River Canyon, fifteen-hundred feet deep, a rugged wound in sedimentary rocks that twisted like a partially unraveled intestine, cutting across the body of South West Africa.

With packs on, we climbed down its steep sides a mile or so into radiant red sunsets and the beautiful dappled colors they produced on the calcareous canyon walls. With quiver trees more than three hundred years old and plants that survived years of drought, it seemed prehistoric. We stopped late in the day at the hot sulfur springs of Ai-Ais for a long, wonderful, skin-tingling pool bath, where we gazed up at the looming cathedral-like walls of the canyon. *There must be spirits in this world,* I thought. *There must be a purpose that man was not allowed to know, that he could only wonder about.* I don't think humans can survive without such hope-inducing thoughts. The canyon seemed a forgotten place, a magical place, a place to forget problems. So we stayed.

A few days later we returned to our Land Rover and continued on over jarring washboard roads. It was a relief to reach Aus and the tarred B4 road that went west into Luderitz. Aus was just a village, a former prisoner-of-war camp for German soldiers after their defeat by the South African forces in 1915. There were rare wild desert-living horses roaming nearby, not indigenous. Some believed they were descendants of German cavalry horses or of horses who had survived a shipwreck. They had knots and scars on their noses and bodies from rubbing them on rocks.

We drove into Luderitz that evening. The next morning Patrick and I, wearing jeans and T-shirts, drove back toward Aus, watching the desert to our right. We were anxious to enter it, as it was the northern boundary of the Prohibited Area, or Diamond Area 1. If we wanted to find the gemstones that lay hidden under the sand, we would have to trespass in an area patrolled by men and dogs. We knew that trespassers were dealt with roughly.

"Look for a place we can break away from the road and not leave tracks in the sand walls," Patrick told me. His right hand rested on essentially featureless maps in between us.

"You have your compass and a spare?"

"Sure. You know not many people have explored the Namib Desert," he added a few miles farther on, where we eased the Land Rover onto a rocky patch next to the road and into the desert. We expected the wind to sweep away what few tracks we left.

"The Prohibited Area, Patrick," I said unnecessarily while looking at the empty road behind us.

It was still morning; the sand was cool and the air spaces hadn't yet expanded between the grains, increasing the chances of getting stuck. We had reduced the air pressure in our tires for more control.

We continued to drive south into the Prohibited Area for about fifty miles and then we planned to turn west, relying on our compasses, toward the coast where we would spend a week or so.

We went slowly, looking for evidence of prospectors or security guards. The Rover reached the top of a gentle sand swell. Below us a small herd of more than a dozen gemsbok grazed fewer than four hundred yards away. Their rare presence and magnificence startled us, just as we, no doubt, startled these creatures of the desert.

Each horse-size body was a reddish gray, and their head and legs were brilliantly scored in black and white. Most distinctive, even from a distance, were the elegant backward-slanting horns, a yard long, tapering symmetrically from the stout ribbed base to lance-shaped points. If I were to toss a pebble into the air in front of a gemsbok, the animal could deflect it with a single flick of the horns. These beautiful creatures existed in the same torrid environment as the scaly cold-blooded reptiles that basked in the sun and required little to eat and drink.

Dropping over the sand swell, we picked up speed on the level plain until we were racing across the open desert next to the fleeing herd. The windshield of the Rover was levered open at the bottom; the forward air ducts were extended outward and the side windows slid open. Hot air streamed through the apertures, and our camping equipment bounced on the metal floor. Patrick gripped the wheel firmly and depressed the accelerator while I braced myself against the dashboard as we enjoyed a close-up view of the timeless beauty of the herd's instinctive flight for survival, panicked but precisely orderly.

Patrick had observed small herds such as this during the course of his fieldwork and other activities. He kept track of their habits and locations. Any change in a herd's behavior might indicate the presence of some element alien to desert life—a guerrilla SWAPO band from the north, for example, or a stray lion or leopard, or a patrol of diamond company police. If it got too quiet, if there were too few animals, Patrick would know something unusual was taking place.

I noticed that one of the females stood out from the rest. She had a small bare patch on her side, as if she had rubbed it on a rock until the fur had worn off, and one of her horns was slightly longer than the other. Her long-legged calf ran gallantly by her side, somehow managing to keep up. But then The Lady, as we called her, began to fall slightly off the pace and lag behind the herd. Her calf also slowed down and seemed confused. Instinct must have told it to stay with its mother but also to survive. Just old enough to run, it could make a distinction if she fell too far behind.

Not wanting to tire the animals further or deplete their water stores unnecessarily, Patrick veered away, and we headed for a depression where we could drop out of sight.

"They have a special adaptation in their nose to help them cool the blood that flows to their brain," Patrick said. "They have adapted to the desert, but even so they will exhaust themselves after such a burst of speed."

We stopped the vehicle, got out, and, carrying binoculars, walked to a slightly elevated ridge where we lay on our stomachs and watched the herd. Sweat evaporated instantly in the dry heat, while my pulse rate showed my heart was beating faster as the temperature increased. I could be burning up calories proportionally faster than the gemsbok. There were no fat men living in the desert.

The air was crystal clear, and from our vantage point we could observe the herd for miles. They soon stopped their run and gathered. We saw one squat in a graceless manner, staking the herd's territory. It was characteristically depositing its droppings in a single mound that would retain wetness and scent for as long as possible.

Intrigued, we saw The Lady, on shaky legs, and her calf catch up. She was obviously sick, probably from one of the common parasite or tick infections. Nothing else could have slowed her in the flight, especially with a calf to protect.

A few days later, our explorations took us to the vicinity of a watering hole about ten miles from where we had seen the gemsbok. It was a pool of muddy water not more than fifteen feet wide; but if the desert had a heart that pumped its lifeblood, this was it. We planned to stop, first to look for tracks, lion or human, and then fill a couple of *jerry cans*—gasoline cans—with the muddy water for cooling ourselves and washing. Before we approached, we paused to view the hole from a distance with binoculars to see if there was anything or anyone around we didn't want to encounter. In the daytime it should have been deserted. Animals mostly visited at night. This time we noticed something strange—an animal lying in the mud next to the still water.

"It looks like a gemsbok," Patrick said.

"Is it dead, do you think?"

"No, it's weakly trying to stand up."

We approached on foot. As soon as the animal sensed our presence it scrambled frantically to get up but couldn't. Of course, it was The Lady.

Just as a sick human needs water, so does an animal. She must have been burning with fever. Desperate strength and determination, the kind that come just before the end of life, had brought her to the watering hole. Using that same resolve, she had probably shooed her calf away from her and into the herd, teaching it one final lesson.

I stayed back a short distance, dropped to one knee, and watched without moving while Patrick got a shovel from the Land Rover. He scooped it full of muddy water and carefully slid it next to her snout. He talked to her gently for a few moments before slowly backing away. He intended it to be her last meal, so to speak, but she wouldn't drink.

"Do you want to do it, George, or shall I?" Patrick asked.

"We're in your territory."

"Damn."

He went to the Land Rover again for his geology pick—neither of us ever carried a gun. With several deft movements, he twisted her head and stood on her rapier horns. As sick and weak as she was, a false move and she could have skewered him. Beneath long and curved eyelashes, her huge eyes, glistening with tears, watched him as he placed the blunt end of the pick against her windpipe.

"I can't do it," he said. "Not with her staring at me."

That evening we camped about a half mile from the watering hole. The desert can be so black at night that if it weren't for the stars in the sky, one might think the world had ended. But that evening the stillness was disturbed by a hungry wail. Something was coming. A jackal? How do they always know when another creature is helpless?

At about midnight, as Patrick and I sat talking in the dark, a high-pitched, hysterical shrieking of mirthless laughter started up, rising and falling, sniveling and howling. It was like the haunting laughter of a lunatic, the hellish chorus sound of the spotted hyena pack. It made me go cold inside. Their calls inspired local people to think of them as knowing and sly, imitating human voices as they did. It meant the jackals weren't finished yet. The hyena, a coward scavenger, waits and eats what is left, but as it waits, it sulks and shrieks. If all the meat has been eaten, the hyena will crack open the bones with its massive premolars to get the marrow. The creature is designed to eat everything.

We were silent for a long while before I asked Patrick how he was doing.

"Okay, I guess. But I should have tried harder."

"She went in the ways of the bush," I said, trying to comfort us both. We had another drink, not enjoying it.

Finally, we fell into a light doze, hearing the last of the series of grunts and moans before the hyenas went loping away with their odd high shoulders rising and falling. We couldn't allow ourselves to sleep deeply. The desert was too active that night.

At first light, circling vultures appeared and disappeared in the sky. This meant the packs had indeed eaten their fill and moved on. The birds would then land like giant black flies and rip at the kill.

When we went back to the watering hole later that morning, The Lady was less than a carcass. Three-quarters of her had been removed overnight. Her back was broken and bent upward. A few jagged and bitten-off ribs protruded at odd angles.

The packs normally devour the hindquarters first, the tender parts, and then the rest of the flesh. The birds had taken whatever soft portions were left—the eyes, nose, and entrails. Now the ants were marching in for any remnants. The space around was trampled flat and strewn with bits of bone and flesh and loose vulture feathers. The sun had already sucked up any fluids the beasts and birds had missed. About the only thing left were the horns.

Patrick and I took them.

Chapter 12

Black Wall of Fear

In Africa incidents manipulate life, if not actually threaten it. They often happen frequently and are of an unusual nature. One hot day Patrick and I faced one.

We spent another week in the Prohibited Area, turning over sand in our search for prehistoric streambeds containing diamonds. We dug up an old Bushman's grave, thinking the owner might have collected precious stones and found the remains of several others. We endured a ferocious sandstorm that could have completely obliterated us.

Late one afternoon the *thump-thump* sound of a helicopter shattered the desert's silence. Fortunately, we were parked between massive boulders at the base of a small Proterozoic ridge and were quickly able to cover the Rover with our sand-colored tarp. The blazing sun merged our presence into the rocks as we squatted under the tarp picking through the pieces of glass-like pebbles we'd collected along the way. "Time to go, Patrick?"

He didn't answer immediately; he seemed to be calculating. "Yeah," he said shaking his head. "They might have found our tracks entering the area from the tar road. But that sandstorm we encountered would have obliterated all others. We can come back another time as long as we keep silent about what we found."

Once out of the Prohibited Area and driving toward Luderitz, I thought again about my problem. "I'll go back to Cape Town and face what will probably be a most uncomfortable showdown with Afrikaner Joost," I said out loud and more to myself.

"He's probably just as concerned about your results," Patrick answered.

"Naw, he sees no threat from me. I'm a nothing. The possibility of his being wrong isn't a consideration for him. He's shrugging me off."

"Maybe he's gone funny in the head. Since he's come to the university, he thinks he's one of the boys."

I rolled my eyes. "You know, it's not as though I'm trying to find the Northwest Passage. I just want to finish my dissertation. My fear is they will stop my funding before I can use the electron microprobe to analyze my mineral assemblages. I suppose selling a few diamonds will help."

"Yeah."

"Anyway," I said, "I'm playing for time while I'm subsidized by an organization associated with the Afrikaner government. That means I'm on an academic *and* political tightrope. If I lose my funding, I'll have to leave." I glanced at Patrick, thinking how much I would miss our friendship and adventures.

"Let's spend some time sightseeing around here, try to clear your mind of all this," he replied. "Then we can move on to my area. As we discussed in Cape Town, the rocks there strike north-south, as yours do in your coastal area. Maybe they are linked by the same later deformational event?"

As we pulled into Luderitz and drove around a bit, I saw it had stayed an isolated throwback to the days when Germany colonized South West Africa. Now, in the 1970s, the town remained much as it had always been. Churches, gathering places with German beer, and bakeries offering German pastries lined its tidy streets, and German architecture stood out: a train station, several hotels, and homes. Another oddity was that the official first language of the country is English rather than a local dialect, German, or Afrikaans. It seemed the local native people didn't have any problem peacefully accepting the influence of their Western past while moving boldly into the future. According to my friends, it was the opposite in Rhodesia and Mozambique.

Upon our arrival, Patrick visited the Anglo-American Foundation office there, which financially subsidized his studies. He was still their employee. My friend knew a lot about the area.

"Diamonds were discovered here, in South West Africa in the early 1900s. Not far away. Over there," he pointed offshore, "is a floating, sand-sucking barge dredging for them. They hope a few diamonds are mixed in with the sand."

"We know that feeling," I said.

"The diamonds here were thought to be from huge volcanic pipes inland, like the Kimberley pipe in South Africa. They were also carried here by ancient rivers and the ocean currents. They brought a diamond rush of some fifty thousand people from all over South Africa and elsewhere."

The coastline to the south of Luderitz, stretching to the Orange River, which is adjacent to the diamond-bearing Prohibited Area, is called the Forbidden or Prohibited Coast. To the north of Luderitz, stretching to the Angola border, the foggy, rocky, sandy coastline is called the Skeleton Coast. All of it is a graveyard for ships. The early Portuguese sailors called the inland Skeleton Coast the "Sands of Hell"—the point being that once a ship is beached and hardship ensues, the crew goes through hell before they die.

The next morning, we first drove to Dias Point. In 1487, Bartolomeu Dias was the first European to see Luderitz Bay. Staring at the sea and the cold, curling-finger waves that snapped out to ensnarl all who dared to enter, I tried to imagine the brave, inquisitive men who sailed over these treacherous waters in small wooden ships around Africa to India and then to China. Those were the times when some thought the earth was flat and that ships could sail off its edge and be eaten by some monster. Those early sailors had untamed audacity.

Five miles inland or so from Luderitz, the ghost town Kolmanskop rose from the sand. Similar to many places in this portion of Africa, the now-extinct settlement was built to accommodate a rush of diamond diggers. Its name, like Vioolsdrif's, was taken from a person—in this case Coleman, whose ox wagon became stuck in the sand there. Simple explanations often stood the test of time.

Sliding and sinking in the drifting sand, we began examining the remains of the town. A two-story home that belonged to a mine manager stood out. Numerous German-style homes with brightly painted walls, a school, a surprisingly extravagant and well-built ballroom theater, and numerous other buildings popped out of a sea of sand. All were well preserved due to the dry climate, but the pelting wind sandblasted everything as it entered absent windows and doors and then funneled across rooms. One smooth sand drift had swept across children's desks, burying them. We walked back and forth on the fairly large theater stage; the faded curtains were still hanging, the props still in the wings. Empty seats faced us, reserved for a ghost audience. It was a deceptive representative of what life once was. During the town's pinnacle in the 1920s, the nearby sands had yielded over five million carats of diamonds in the first six years of its existence. Then the town, with its comparatively small population, had the highest per-capita income in the world. The stark abandonment reminded me of The Oasis and the lopsided swing. Some things in life are so final.

Kolmanskop was finally abandoned in 1954. Many rushed to another rich deposit of diamonds discovered to the south at

Oranjemund on the mouth of the Orange River. Others probably moved to Luderitz. It seemed to me that in Africa memories fade quickly; life is too uncontrolled and chaotic to be burdened with too much memory. Kolmanskop was not only a ghost town but also a tombstone.

We returned to Luderitz and a day later we drove east some hundred and fifty miles toward Keetmanshoop and Patrick's field area. There we bought provisions, including a couple of large steaks. We left the tarred road and entered the perimeter of the Kalahari Desert, driving north until dusk. It was time to camp.

Our frying pan had been lost, so we cooked our meat on a shovel, careful to not burn its handle. Under a clear sky, with cheap red wine in our tin mugs and tender meat on our metal plates, we sat talking. Faraway stars, their names slow in coming, hung like captivating beacons in the unworldly pristine air. The sounds of the desert night stirred our imagination. It is odd how rapidly creatures make their presence known and announce themselves. A lion growled in the far distance and unknown insects screeched around us. Sounds can cast a spell. It was one of those memorable evenings when memories are cemented forever, when sand is better than a mattress.

The next day we were off, walking up a dry stream channel and looking for rocks we thought were deformed by the north-south event that affected the far coastal region of my study area to the south. If that event extended north into Patrick's area, it would add to my theory. The stream led us into an open plain. It was an attention-grabbing place, as there were indeed rocks of interest to both of us. Absorbed with the idea of finding more, Patrick and I kept walking for an hour or so, our eyes watching the ground, taking in every rock.

"What the hell is that?" he said with some urgency, looking up into the distance.

We stared; it was a mass of black moving in our direction and not so far away. They were animals: well more than a hundred Cape buffalo walking slowly with their heads down, nibbling on whatever they could find. A wall of them, raising clouds of dust. The lion will growl before he charges. The buffalo simply comes forward in a silent charge. Nothing stops him except death.

"The buffalo is the most dangerous of the big five. They will ambush hunters," Patrick added.

"Oh shit."

"They're coming directly at us," Patrick said. "Look at them, marvelous massive bulls with mountainous bosses of black horn. Some of them must measure thirty-five, forty-five inches or more around the curve from point to point. We should get out of the way."

"Brilliant. And where do we go, for Christ's sake? There are plains all around us."

"The tree."

He pointed. I looked. An acacia tree stood alone in the plain. It wasn't very thick, maybe a foot in diameter, but it wasn't far away either. *Perhaps it was just strong enough to hold us about ten feet off the ground?*

"Good Lord," I mumbled.

"You can certainly ask Him, as this *will* take a certain amount of faith. Let's move slowly. No impulsive movements. No sprinting. If that mass gets excited, we will get horns up our asses."

We reached the tree without panic. Being taller, I climbed first, as I could reach up farther since the lower branches were lacking.

"Ah, shit," I mumbled as I got pricked by its long thorns. I wrapped myself around a limb and reached down for Patrick. I grunted, trying to lift him.

"Come on now!" he complained. "They're getting closer!"

"If you didn't eat so fucking many potatoes—"

"Do it, man!" he said frantically, swinging up his hanging legs, holding on to me.

"Oh shit," I groaned as I lifted him up. He clawed his way over me.

"Well, now," he proclaimed. "We've a good view."

Very clearly the phalanx of Cape buffalo was munching its way directly toward our tree. Rooted on stocky legs, the beasts are the size of American buffalo. Closer now, we could see their bent heads—the dimension of champion pumpkins—as they held their massive curved horns close to the ground. From time to time muzzles, swinging back and forth, lifted in our direction. In the thick folds of their throats hung dangling bunches of ticks. One of them drooled a thick string of saliva. Their poor eyesight focused on the available blades of grass. As for not seeing us, it made no difference; the wall was coming our way. They hadn't smelled us yet, either. It was plain that if they wanted to, they could easily push our tree over and trample us.

"Oh hell," Patrick said. "Let's not do anything to put them in a bad mood."

"Good thinking."

"Nothing more we can do. We've got nowhere to go. As long as they stay calm . . ."

"Us, too," I replied, seeing Patrick's dangling foot tapping the air.

They kept coming. It was captivating and scary. I looked down at the thin limbs we relied on. We watched in silence. Oxpecker birds rode like jockeys on the animals, stuffing themselves with ticks and whatever else infested the buffalo's thick, dusty hide and fur.

Closer . . . The wind was in our direction; we smelled their reeking bovine odor made all the worse by the heat. Above us the African sky was seething gold. The buffalo were fewer than one hundred feet away when one of the birds made a sharp sound.

The whole herd paused, standing massive and black. They were attentive and ready; they froze as one, their noses held high.

"Don't move," Patrick whispered.

"I don't think instruction is necessary." Another tweet sound from the bird. "I wonder what that shit bird is telling them."

"Look. They're turning," Patrick whispered.

It was true. Like marching men, they pivoted as one to their right. With a couple of fucking tweet-tweets, the bird had directed them to change course.

"I'll bet the buffalo couldn't see us even this close," Patrick whispered. "Those birds have better vision. They figured going around us was a better idea, so they could stay away from problems and continue to eat ticks. It was a field-smart decision."

We sat in our tree in silence, relieved and waiting until the buffalo ambled off, in no hurry while the birds patrolled their black backs in the warm sun.

"They could have trampled us into two spots of bones and blood," Patrick said. "I saw an angry one once."

"Let's get out of here."

We climbed down, looked at the parched tree that saved us, and urinated on it.

I took a piece of sausage from my pocket, cut it in half with my clasp knife, and handed a piece to Patrick.

"The African bush, it's fascinating and terrifying," he said, taking it.

"Yeah. Let's look for rocks, Patrick."

We crossed the herd's hoofprints. They were the size of soup plates.

Chapter 13

Sacrificed

Patrick and I drove back to Cape Town with me wondering again about my situation. It was infecting my mind like a disease. I had to face Afrikaner Joost. Why did this geological discrepancy have to happen to *my* project? It seemed an unjust allocation of responsibility for a student to have to deal with.

I never wanted to stand out, let alone as the bad guy. I did not want to think about grossly manipulating the data I'd found since I began my project—over the past year and a half—simply so I could agree with a confused professor. I couldn't get it off my mind. However impossible the idea was, hell—there was *no* middle ground.

I'd gone over and over the data since I started questioning Joost's conclusion some ten months ago and began living with the resulting confusion and anger. I had no choice now but to take the first major step into the fire, even though I unfortunately didn't have all the electron microprobe data I needed. I'd have to take a chance that I'd be able to explain my findings and be believed. Perhaps I'd delayed too long. But working in the field and building a geological story of what happened to the earth *a billion years ago* takes time. That's why four years of total study are advocated to prepare a PhD dissertation.

I dreaded the confrontation, but the time had come. I presented my current conclusions to Afrikaner Joost.

He smirked and said—yet again—he was sure he was right. He didn't even ask to see my evidence. He brushed me off, smiling as if I were a child.

I spoke to my advisor, Professor NASA. Again, I left no room for exceptions; I showed him on a map what I'd found. From his answer, it seemed he either didn't understand metamorphic petrology or he *didn't want to get involved.*

How could that be? The whole thing was a fucking circus. He said he would talk to Joost. But it seemed that Joost had already spoken to him and convinced him I was wrong. It didn't feel like a topic open for discussion. It was scary and confusing.

My private office was relocated to a large community room. It was a move to degrade me. And it worked: I was dropping down the ladder rung by humiliating rung.

From whispering to open discussion, everyone in the department became aware of my situation. I was walking the plank. Department personnel stayed visible friends with Afrikaner Joost, who kept a sociable know-all smile on his face. He was the guy with the money; they needed him and Professor NASA supported him. He *knew* that. What would be their next move?

"You are to present your findings to the whole department," my advisor said. "I want you to take this presentation seriously." Beyond this, he did not offer one scrap of help—not a word of guidance. I reluctantly concluded later that he really didn't want me to succeed. I was being set up. I listened to him, though, because his academic accomplishments were formidable. He'd been a scientist at NASA, for Pete's sake, which put him into the legend category. Who wouldn't listen to him?

I had never defended something as conflicting and meaningful as this before. I had never presented a scientific lecture that was more than a basic report. The one person who seemed to be willing to help me—Peter F., the only metamorphic petrologist in the department, who had offered to replace NASA as my advisor—suddenly left for a new university position in his homeland of Australia. He was a good man with a conscience. He sympathized with what I was saying and was apparently willing to go against Joost. However, now that he was gone, that option was obsolete, and I didn't think things could get much worse. Unfortunately, that's exactly what happened: They got *terribly* worse.

I reminded myself that to support my field findings in a manner that would force others to take notice, I needed to subject my mineral specimens to microprobe analysis and then apply certain equations to the resulting chemical analyses. Those methods had been denied me.

The people who would be in the audience were mostly laboratory and geochemistry men, not field men. They would understand conclusions based on laboratory data; it would sway them. Unfortunately, all I had was field evidence, the ringing of rocks and thin sections, so to speak. In science, if you're going to make statements, you have to prove them.

Lecture day came. Deep down, I believed right would conquer. Looking back, I can't believe I was so naive.

About thirty people—virtually all the department's staff—filed into the small room and took their seats in the bleachers. All were invited, I later understood, to witness my downfall. My friends Patrick and Catina were also there. Professor NASA and Joost were in the third raised row—that put them eye-to-eye with me.

Professor NASA and Afrikaner Joost whispered together and smiled freely while I spoke, as if they were at some cheap movie. I showed slides of thin sections, what rock analysis I had, and discussed relevant metamorphic zonation. But I was unnerved, apparently unable to make my findings clear.

Faces stared at me in silence. The others listened, trying to follow my presentation, trying to grasp that what I was saying was the *opposite* of what a well-known professor—who had the support of Professor NASA—had proclaimed as fact. Their faces registered disbelief. *How could this student be right*? It was as though I was committing some kind of travesty. Although I suspect some of the audience wanted to agree with me, I hadn't presented enough evidence for them to climb over that enormous wall and come over to my side.

I felt I'd made an ass of myself. Joost and Professor NASA sat smirking and whispering to the very end. Right did not win; I had been sacrificed. Perhaps I should have acted sooner, but had I confronted Joost earlier, what happened next would have happened sooner—and I would have had even less time to collect evidence supporting my dissertation. I had to painfully play for time to get this far.

"Your findings don't prove anything," Joost said a day later when he passed me in a hall. He spoke in a monotone and wore his smug smile. His eyes, the windows to that man's soul, were blank. Phony. To me he was sly, a snake—he had no substance. And he was winning. I wondered if anyone really liked him.

But that wasn't the point. There he was, and he wasn't worried. He could be cool, he knew the extent of his power, and, yes, he had won. "*You are wasting research money*," he continued, smirking.

A week later, a Friday afternoon, Professor NASA, my "advisor," called me into his office and told me my funding would be cut off *in two months*. I could no longer use the Land Rover. That meant I was tossed out two years into my study. He said this in an understanding, sympathetic way, as he held all the cards and could play magnanimously. He, like Afrikaner Joost, had gotten what he wanted. Professor NASA was used to getting what he wanted.

Even though the possibility always loomed that I would be told to leave the university, I couldn't believe it. When it actually happened, I was shocked to the point where I could barely stand. I went home in a daze. When I finally was able to sleep, I dreamed of failure. Following dreams demands a price, no question about that. I told myself that in the end I would be stronger.

Somewhere I had read that when all is gone, a store of hidden strength and resolve comes forth. Some men call on it; some men do not. What would I do?

Should I give up?

No, I decided, once the shock began to wear off. I had to stop being naive. It was time to punch them back—to stop hoping for the best and play it their way. But the clash would be very difficult. Did I want my degree that bad? *Yeah, I did.*

Their premise was flawed and their treatment of me was wrong. I was on the bottom rung, so what did I have to lose? I had two months left before the final payment of my research grant. I'd finished 80 percent of my fieldwork. It was time to attack Afrikaner Joost's and Professor NASA's glory.

To do that, I would need enormous staying power. I would have to eat buckets of shit for the next two months, until hopefully I could think straight and dig my way out of the terrible situation I was in. I stopped discussing the subject with others and painfully absorbed all the pitiful looks I was given as I carried on my work in the PRU as best as I could with the data I had. I talked only to Patrick and Catina, who both understood.

Now that I unequivocally understood I would no longer be associated with UCT, that I was indeed thrown out and no one would help me there, a plan crystallized. I would have to seek help outside UCT from someone who knew South African geology.

This wasn't easy, as I was a foreigner in the country and the only people I knew were associated with UCT. Also, I was going against an Afrikaner. This increased the odds against me, and I realized it was possible my pushing back could make the situation even worse.

* * *

I remembered a friend of mine, Dr. Martin Jackson. I'd met him when I first arrived at the PRU; he was in his final year of PhD studies then. Now he lectured at the competitive University of Natal, on the opposite—east—coast of South Africa, at Pietermaritzburg. It was not far from the large city of Durban. Martin specialized in metamorphic petrology and structure geology—just the combination I needed.

I called Martin and explained my findings. He said he understood and was *curious*. Curiosity in a scientist is magic. He was also from Rhodesia, which meant he was not unduly influenced by Afrikaner power. Had my luck changed? His own PRU research area, although much farther afield from mine, had, he admitted, made him wonder then about Joost's findings. As I was no longer associated with any university, he suggested I apply to where he lectured. He offered to be my advisor if I was accepted.

But, I wondered, *do I have enough data yet? Will this work?*

I needed another trip to the field to examine one remaining area. I drove my old car. I hoped it wouldn't break down, especially when I loaded rocks in it. I lived from the trunk like a gypsy. I didn't tell anyone except Patrick and Catina I was going. I was wary of what Joost would do if he found out I was not going to give up. Could he use his considerable influence to try to block my admission to the University of Natal? I believed anything was possible. After all, he'd stopped me so far. So, secrecy was essential; secrecy was part of the battle plan to outflank him.

I spent what money I had on gas, ate bread, and drank coffee. The African insects buzzed and the whole earth was my bed. Tattered clouds, like my life, floated across the blue sky. In the morning, heat waves rose. The billion-year-old rocks had talked to me, and I owed it to them to tell their story accurately.

It felt good to get out of the geology department building. I didn't belong there. Some people were embarrassed by my presence, not sure if I really was wrong and perhaps guiltily wondered if they should take my side, while others preferred to see me as someone who'd falsely bucked the system they felt quite comfortable in.

My old car didn't travel well with its weak brakes and its risky steering. After three weeks I would return to UCT in as ragged condition as that car.

While in the field, I met a new PRU geology student named Alan, who kindly helped me collect some rocks. His field area overlapped a portion of mine. I explained Joost's findings were incorrect, thinking it would help him avoid presenting erroneous information in his dissertation. He replied honestly, "I want my degree," and that Joost was responsible for bringing him to the PRU. In other words, Joost must have told him he expected Alan to support his findings.

Maybe Alan didn't believe me, but I think he just wanted to take the easy way out. On that note, it fascinated me that in the mornings when he woke, he would lazily walk to the very edge of the small awning in front of his trailer and pee rather than taking a few extra sanitary steps. Unfortunately, Alan reported to Joost about where I'd been, but I was able to keep them in the dark for almost a month. For my second and final month I continued working away in polite silence. There was a lot of tension. The whole department wondered why I didn't give up. "What is George going to do?" they quizzed Catina and Patrick again and again, all trying to find out and rat to Joost and Professor NASA. My two friends said nothing.

Within days of my return, Professor NASA strangely made the geology department's complex microprobe analyzer equipment available to me for a four-day holiday weekend. I didn't understand why and didn't ask. Perhaps he had a change of heart? No, I couldn't let myself be sucked into that false rubbish anymore. More than likely, he and others suddenly wanted to know about my findings and then influence anything I might present. Plus, they likely didn't want anyone to say I didn't get a fair chance—they were covering their asses.

No matter; I needed the data and played the game back at them, struggling to smile and say thank you. They saw only what they wanted to see. The horns of David and Goliath were locked.

When the time came and the machines became available, I hardly slept for those four solid days. I never left the building. Catina, a remarkable woman, shuffled food to me and lots of coffee. Day and night I analyzed as many specimens as I could. Twenty hours a day I worked as only the young can do. I slept on the floor. When the printouts came rolling out of the machine, I watched with tired eyes and was ecstatic; the data supported all I had said.

To a scientist, facts are strength. My God, *I was right.* It was the first time in a long while I truly smiled. Catina said she finally saw life back in my eyes. I would have to go through the data in detail to be doubly certain, however. In the meantime, I told no one.

The end of the month came. In silence I packed my documents, books, and selected rocks and retreated with them from the university to the nearby apartment I'd rented with Patrick two years earlier.

It's true there is no middle way in Africa, I thought. *It's either death or life, lose or win.*

Was it their turn to lose?

Chapter 14

The Beginning of Environmental Science

Looking around my apartment, I recalled the past two years of my life: a small wagon wheel wired as an overhead light, which I had taken from The Oasis. A dozen spent mortar shells from the 1915 South African–German war that Patrick and I had found in the Prohibited Area. A hundred-year-old black cast-iron Afrikaner cooking pot I'd removed from the sands in South West Africa.

Special rocks Patrick and I had collected adorned our shelves, such as an eighteen-inch-long, egg-shaped nodule of beautiful green malachite; another with black tourmaline crystal prisms embedded in it rested on the floor. There was blue copper ore, green and blue diamond pipe rock, and a two-foot-long, six-inch-wide slab of the extraordinary blue lace agate sitting on a bookcase.

I reminisced then, remembering another challenge. I was seventeen, it was the 1960s, and I'd hitchhiked across the United States from Buffalo to Chicago and then north to the Dakotas and west across the Badlands to Seattle, Washington. From there I went north to Mile Zero on the dirt and gravel Alaskan Highway and then west to Haines Junction (historic mile 1016) in Yukon Territory, Canada. My goal was Juneau, the capital of Alaska, where my brother worked.

As there were no roads into Juneau, I took a ferryboat to get there. I returned via the length of California to Bakersfield, went east on Route 66 through the Mohave Desert, and then on to the Great Plains of Oklahoma, to Chicago, and home to Buffalo. I carried two large double-sided cards. On the first I'd printed "North" on one side and "South" on the other; the second card had "East" and "West" on it. I stuck the signs between two ropes that bound my old suitcase together. I slept in hedges, cornfields, and deep within mosquito-laden forests. If I could do that, I should be able to face the homelands of Natal.

In the meantime, I had to get there and that meant considering: *What would come next?*

The answer came soon enough.

Several weeks passed while I worked in my apartment, compiling the analytical data I'd collected. On Sundays I roasted a chicken for Patrick and Catina. Joining us one afternoon was an untidy Australian geology student named Marcus. Bearded and with dull red hair, he wore loose-fitting, dreary gray clothing with a burnished patch on his knee. We were the same; our economic status precluded pretensions. We'd met Marcus about a year before at a local student bar when he visited Cape Town on vacation, and we kept in contact. He was from a farm in the outback and now took graduate geology courses at the University of the Witwatersrand in Johannesburg part-time.

I brought the roasted chicken to the table, which sat under a tattered wall map displaying spattered grease spots on the African continent.

"It would be nice to travel a bit into Africa," Catina said wistfully.

"Now seems a good time," Marcus said.

Our eyes turned upward to the map. It was a standard Mercator projection, most accurate near the equator, which we were not so far from. Often Patrick and I had gazed at it, musing about various sights to be seen. I stood and used a thighbone to point with as we discussed locations to our immediate north. A few minutes later we made the decision to drive northwest to the Limpopo River. From there we would enter Botswana and cross a portion of the enormous Kalahari Desert on our way to the great Okavango Delta. Then, we'd turn east to Victoria Falls and the Zambezi River and enter Rhodesia. Probably in Bulawayo—I saw an airport key on the map there—I would get a plane for Nairobi, Kenya, and later fly home to see my family in New York. The others would travel back to South Africa. At that point we'd have traveled about five weeks.

"That's about two thousand miles to Bulawayo over rough roads," Catina said, holding a wing delicately. We knew she had to stay behind and work. "You know they are having a civil war in Rhodesia?"

Patrick looked up again from his potatoes (an escaped piece had lodged in his beard), adding, "There's also gold in a stream there, according to George."

"Well," I said, "the white hunter showed me a nugget. Anyway, we can decide on that part later."

We roughed out the whole trip in two hours or so. It would take us into an Africa we hadn't seen before, and that meant adventure. The major cost would be for gas, which was cheap then. We would sleep on the ground and eat available local food. Standing, we made a three-way handshake on it. "If all goes well," I added, "I'll borrow plane fare from my family in the US and then fly back here, collect my car, and drive to Natal to continue my studies."

A week later, I said a final goodbye to Portuguese Catina. Emotions don't travel well. We knew there was just no other way. We were students full of life, and these past two years had been stressful. I was finished here. I had to face that. In the end, most people were probably glad to see me go.

* * *

It was a sensitive day, a traveler's day. Patrick and I started off in my old car on the way to Johannesburg, where we would meet Marcus and then drive his Land Rover north.

We moved slowly, limiting strain on the car and conserving gas money. I had forgotten to tell Patrick that the gas gauge didn't work well. In the first fifteen miles, the needle dropped from a full tank to three-quarters. I could almost watch it move toward empty.

"I've never seen anything like it!" Patrick gasped as he focused wide eyes on the gauge.

"Almost time for gas," I replied with a straight face.

"Bejesus, we've only gone a few miles!"

"It's an old car," I said.

"'It's swallowing gas like a drain! We'll be broke in fifty miles and pushing it all the way to Jo'berg."

"Aw, it's the gauge. Look." I rapped it with my knuckles and sent it toward the full side.

We drove northeast into the huge flatlands of the Free State. It had a no-ending feeling, a bit like being pushed in a rowboat into the ocean. Sheep milled together on the road as sheep do. Signs said *Bowen se melk en produkte* in Afrikaans: "Milk for sale." Farther on, we passed small, baked-white houses seemingly out of a fairy tale in *dorps*—small settlements—not noted on our map. People lived around watering holes as isolated as a pinprick on a sheet. By the time we thought about stopping, we were well past them.

Our car moved slowly but steadily, mile by mile. The road was thin, isolated, and went straight to the horizon. I had traveled such roads before; they were like following a string to nowhere.

Marcus was waiting for us in Johannesburg. We left my old car in a university parking lot and drove toward the Botswana border at Ramotswa. It was one of many out-of-the-way crossings over the

Limpopo River, which divides South Africa and Botswana. Some of the crossings are closed during periods of high water because the bridges are underwater.

I'd been there before, on a short geology field trip, marveling at the imposing pinkish-gray baobab trees, smooth, shiny, and unchanged since prehistoric times—some living for as many as two thousand years. The early farmers said if the tree failed to bloom in October, the harvest would be meager. Legend says that a python lived in one gigantic baobab; it granted fine crops and good hunting when worshipped. The first white hunter arriving there killed it, and a disastrous crop and other consequences soon followed. I don't know about the legend, but on another occasion, while walking along and reading a map in that area, I tripped over a python sunning itself. I sprinted away and he continued his nap.

Armed border guards made us unload all of our goods for inspection. Our rough field appearance, accents, long hair, and the beards on Patrick and Marcus prompted questions concerning our objectives.

Were we mercenaries? they asked.

"Don't look so wild, Marcus," I whispered, noting his neglected hair had escaped in untidy waves from each side of his baseball cap.

"What the hell do you think *you* look like?" he shot back.

Crossing into Botswana was emotional for me; once I left South Africa, I could not return. Unlike my friends' documentation, my entrance visa would expire, as UCT would not extend it. Once over the line I would enter a third world country and be subject to its laws.

The white border guard looked at me. "Are you sure you want to leave our country? One step over the line," he said as if reading my mind, "and you can't return."

My friends looked at me. The question hung in the air like a tear in an eye. It *really* was a scary decision.

I scanned ahead, across the border: black guards, a few shacks, and a road into the trees. What then? I remembered the remote

Alaskan Highway that I had walked years before as I crossed North America. It drove straight to an enormous forest and never really emerged. Some things are like that, one step more and nothing is the same again.

"Talk about reaching the Rubicon," I said, drawing on courage. With that, I sighed, looked at the guard, and stepped daringly over the broad line into Botswana as they all watched me.

"That was kind of theatrical, George," Patrick said quietly as he caught up with me while Marcus drove across the border.

"It was also creepy," I replied as I tried to reach whatever was biting me on my back.

"Well, you seem to handle it okay."

"I seem to be getting tested too much lately," I mumbled, looking at Botswana while stopping to rub my back against their flagpole.

Via a combination of asphalt and sandy streets, we reached Botswana's capital, Gaborone, not far from the border. The town's center was ill defined, as the buildings were mostly one story, with later additions crawling in various directions. They were spaced like an old man's teeth—with plenty of gaps. Paths led between the spaces, and people parked haphazardly at each building's entrance. In the middle of the city, we parked the Land Rover on a sand rise and partially under a scraggly acacia tree.

Dust rose in the hot air as people shuffled their feet and socialized. Tribal women dressed in bold flowing gowns with large turbans of bright material twisted around their heads disappeared into buildings and around corners. Many in the streets balanced on their heads what Western people would carry in their hands. Men in sandals had cloth wraparounds, off-the-shoulder rough cloth, and dusty robes; they carried sticks and spears as they wandered from one thorny acacia to another. Mixed in with this were Western jeans, skirts, T-shirts, dress shirts, hats, and shoes of all kinds.

Abundant wild trees grew at will, or at least until someone

decided they were in the way. There were bleating goats tied to most of them with long rope leads. Few leaves existed below the level the animals could reach while standing on their hind legs. Bare spider-like branches remained. Weary donkeys slept waiting for their owners, their tails continually flicking at flies. Thin dogs lurked in the shadows. Mostly, they slept and scratched.

Gaborone was a friendly, loose kind of place. It was the capital of an African country run by rural Africans, who still had wilderness in their eyes. It was a frontier town in that there was no pollution from progress. It was a town I liked.

Now, forty years later, I understand Gaborone is one of the fastest-growing cities on the continent. It has modern high-rise buildings, a hospital, a five-star hotel, schools, recognizable government offices, and other structures that would be found in a Western city.

At Gaborone, we bought food and fresh items that generally cannot be taken across a border. With a month's supply of goods, we backtracked to Ramotswa and then to Lobatse and began to follow the only road north to Phuduhudu, Kang, and finally Ghanzi, which sat near our goal, the great Okavango Delta.

To get there we had to travel four hundred miles or so of untarred trans-Kalahari road. The red roads of Africa. The Kalahari Desert is a large arid to semiarid sandy area that has some small amounts of rainfall, which means it's not a true desert. It is massive; the bulk of it stretches over three countries, 80 percent of it in Botswana. Surrounding it and extending into other countries is the similar Kalahari Basin. Together they constitute one of the emptiest landscapes on the entire planet and can be seen from space. And we were to enter it in an old Land Rover? Youth is hopeful.

Standing on the roof of our Land Rover and peering through binoculars, I saw the ocher land was flat—a seemingly endless plain, as if it were possible to march directly to the rim of the world. When the hot August winds blow, Kalahari sand can be carried across the whole southern portion of the continent. At the same time, the tracks of Land Rovers like ours can leave impressions that last decades. It was that kind of place; imprints could last a day or months or years. The only landmarks were the occasional odd-looking stunted trees, bushes, and ephemeral rivers. Without the sand road, it would have been easy to lose direction and wander off. It was said even the Bushmen, who were walking compasses, could get lost. They called the Kalahari the Big Dry.

It was so silent. How could something so big be so quiet?

Every so often the sunken, pitted remains of a stream or riverbed came, gouging the road. In the spring or after the rains fell, these channels would fill briefly and flow in some direction until the Kalahari opened its sand mouth and swallowed it all. In one we saw dried wisps of grass and light-green striped *tsamma* melons. When they mature the rind gets tough, protecting the water inside. When the rains come, the rind rots, releasing its seeds. Red cucumbers nearby would rapidly ripen with the coming of rain; a bit of water would be enough to fortify the odd, gnarled acacia tree and mimosa shrub joining them.

In the distance, there seemed to be a huge plate of salt.

As we drove, Patrick gave us a quick language lesson. "The word 'Kalahari' is from the Tswana word *kgala,* meaning 'the great thirst.' In the heat of the summer, there are mirages, which appear to have water. The camel thorn tree grows near salt pans, and that helps promote the oasis idea. This confused people, especially when there were no roads. Years ago, travelers in the dry season often failed to reach the swamps."

He continued. "The Bushmen and a tribe called Bechuana were able to find water by recognizing the exposed withered stalks of tuberous plant roots that store moisture. They knew below them water, in the form of moist sand, could be found. To get that water one of them, normally an old woman dug a hole several feet deep and positioned a ball of grass in its bottom. Sticking a long hollow reed into the ball and covering the hole over with sand with the reed extending to the surface, she laboriously began to suck on the reed. That created a vacuum in the ball, which drew moisture from the surrounding sand. After several minutes, water went up the reed into her mouth. From one side of her mouth a thin, short stick led to a shell from an ostrich egg. A minute or two after that, a trickle of water would begin to run down the short stick into the shell."

"You mean she sucked water up like a pump?" Marcus asked, incredulous.

Patrick nodded. "Drawing air up the reed creates spaces between the sand grains for water to move into. It's the same way an oil well works. Basically, they invented the concept."

As we drove along, we saw in the distance the unmistakable silhouette of a four-foot-long secretary bird striding majestically. In addition to their long legs, which promote a stately walk, their height is accented by a crest of feathers projected from the back of their head. The feathers are reminiscent of the era when secretaries inserted their quill pens behind their ears, hence the name. As we watched from our windows, the bird suddenly ran toward our vehicle, only to stop some hundred feet away. With its wings fully extended, it began what appeared to be a very complicated, fast-moving dance.

"Is it mad?" Marcus asked.

"No," Patrick said. "It's discovered something—likely a snake."

The heat notwithstanding, we stopped the Rover to watch. The bird was using its wings as a shield against the snake's strikes while delivering violent blows with its feet. After about twenty minutes we thought the snake was dead, but we weren't sure. Neither, apparently, was the bird. It began an elaborate procedure of kicking, throwing, pecking, and watching for the slightest movement before it swallowed the reptile whole. Better to be exhausted than bitten in the neck as the snake went down.

As we drove on, the endless sand road passed through a few dusty villages and clusters of huts. It was so hot, the dogs didn't bark. They lay looking at us, too lazy to fully open their eyes. "It's said that at times the centipedes emerging from their holes are roasted alive in the heat," Patrick told us.

We had about a hundred miles to go before reaching the delta.

"Didn't Livingstone first find the delta?" Marcus asked.

"Yeah," I replied from the backseat. "I read that in 1849 or 1850, he traveled from South Africa to this region of Botswana. He and his party first discovered Lake Ngami, which teemed with masses of animals. A few days later, his party reached the waters of the Okavango Delta."

"That was only about one hundred twenty years ago," Marcus replied, he must have taken the same route to get here that we did.

"At five thousand square miles it's considered the largest inland delta in the world," I said. "More than ten thousand animals a day came and drank from the lake. In Livingstone's time they had no fear of humans. They didn't know what guns were."

"The first sign that something is happening ahead in flat terrain like this usually comes from birdlife," Patrick added. "They can be seen from a great distance, especially if the air is not polluted."

If a bird you recognize is attracted to something below, you can generally anticipate what the object is. The classical examples are vultures circling over the dead and dying. They don't waste time on something expected to live. In front of us, soaring and circling high were several specks. Although we couldn't be sure, they were probably large, gaunt storks, which characteristically gyrate in thermals and then glide off. They have a very high tolerance for the presence of (live) humans and are not easily frightened. Oddly enough, they have no voice, but clap their bills instead.

"Lake Ngami must lie below their spirals," Patrick said.

It was the lake, all right, but there was little hint of a hunter's paradise. Natives lived on its shores with their cattle. The men paddled and poled dugouts, setting their nets among abundant birdlife that was also fishing. Dead trees and dark stumps starkly protruded from the calm, smooth water. Marabou storks, the same we had seen in the sky from a distance, stood on their long, pencil-thin legs in the branches of the dead trees, probably watching for elusive frogs and fish below them. Everything was watching for food.

"The skeleton trees look creepy," Marcus muttered.

"Actually," Patrick said, "what happened to them was kind of creepy. After Livingstone's visit, hunters and others came to the lake. That's when the problems began. They abandoned the rafts made of papyrus they'd used to transport themselves down the main channel. The papyrus took root, multiplied, and slowly strangled the channel, causing the lake to dry up. As this happened, the fish became more concentrated and readily available to the ever-increasing numbers of birds who came to feast. The hippo and crocodiles were forced to leave or died trying. In the stagnant pools of water that were left behind as the lake receded, mosquitoes bred, and soon malaria broke out among the baTawana, who raised cattle on the shores of the lake.

"A drought came, and more cattle died from lack of water and food. By 1896, some *thirty-seven* years after Livingstone found the lake teaming with life, it was virtually dead. It was one of the first ecological disasters caused by man in southern Africa. Because of such catastrophes, a study called *environmental science* would later develop."

Patrick continued. "In the early 1950s, water began to enter the dry lake as rainfall increased. It drowned the trees that had subsequently grown. But for some reason the major fluctuations of rainfall were out of phase. In any event, it brought the animals back. Then, in the 1960s, a drought came. Everything apparently died."

"How do you know all that?" Marcus asked.

"I lived in Botswana for four years. One fellow told me the vultures were so well fed, they couldn't fly. In the 1970s, the water came back again. Now the algae eat the cow shit, the fish eat the algae, and birds eat the fish.

"To survive," Patrick concluded, "life adapts. It may not be easy, but the alternative is defeat."

Yeah, I thought, thinking of myself, *I've also eaten shit to survive. Let's hope if I reach Natal, it will be like the water coming back.*

Chapter 15

White Hunter Remembers Livingstone's Folly

Continuing east and moving slowly so as to not make too much dust as we passed the settlement crossroads called Maun, our goal was the Makgadikgadi Pans to the east. We would camp there for the night. The Makgadikgadi is made up of several smaller pans, like the Sowa Pan. *Sowa*, in the language of the Bushmen, means "salt." Overall, the salt-pans region was huge. Nothing lived unless it could move on its own feet in and out of the area, and the rhythm of life didn't change from generation to generation.

Before recorded history, the pans were formed by a lake that covered most of northern Botswana. The water obviously contained a certain amount of salt, and as it evaporated, the salt condensed, making the remaining water increasingly salty until it lay on the ground like snow. Only after the rains come are the pans reborn, virtually overnight, into great, shallow stretches of blue water. Like the Land of Oz, they are almost mystical places where one world becomes another, when a dry pan becomes a lake teeming with millions of animals. The animals sense the transformation and come to drink the water and lick the salt. In the process they also forage on one another.

Earlier we had filled our roof rack with wood from the dead trees of Lake Ngami. That evening on the perimeter of one of the pans, we built the largest and most memorable fire in all my days in southern Africa. It rose higher than six feet and lighted our suntanned Irish, Australian, and American faces as we sat around it on the sand. It was a fire in the middle of blackness, endless blackness rudely ruptured by frantic sparks rocketing outward with unsuppressed energy. It was a fire that wrenched thought and reason from ideas that otherwise stayed in darkness. It was a fire that rivaled thousands of jeweled stars hanging in emptiness for attention. It was a magnificent fire, it was a magnificent evening.

Our Africa was campfires by night, geology and wild things by day.

It was an evening for stories, as huge succulent steaks simmered over glowing yellow-red coals while Patrick's favorite, potatoes, baked beneath them. We smoked cigars and drank our usual cheap dry red wine from battered tin mugs. Patrick reminisced about his life, poverty, and accumulating every penny to attend the University of Dublin. He mentioned a female he invited to a university dance. She accepted and he bought tickets. Then she refused him and attended with another. He'd saved for months for that event and never spoke to her again. Marcus talked about life on an Australian farm, the heat, the emptiness, and the desire he'd had to learn and see something else. They were my friends telling simple stories about simple events that shaped their lives. I listened and spoke, forgetting my problems and disillusionment as our fire launched its rocket flames into the dark sky. I felt young and in love again with dreams and hope.

There were realistic reasons for the large fire other than cooking; it was lion country, and they are afraid of the flames. It should blaze all night. We agreed to wake up in turns to ensure that. In addition,

Patrick placed upright shovels around his head as he crawled in his sleeping bag on the perimeter of light. Marcus and I stared at him.

"Lions often drag people out of their sleeping bags by the head, like removing a hot dog from a bun," he explained matter-of-factly. "The shovels may confuse them."

"Oh, I thought you were just preparing yourself to take a crap in the middle of the night. What about Marcus and me?" Not receiving an answer, I looked around for a way to protect myself. Marcus lay down with his head under the Land Rover, and I decided to stack backpacks around my head so I could at least see the sky. Actually, I felt like a small piece of meat left on an empty white tablecloth haloed by our fire candle. "There's even salt here to make us tastier," I grumbled.

Tiredness overcame us and no one woke up to feed the fire, which, insatiably hungry for food, had reduced itself to a glow of its former self by the time the sun rose. Lingering by the pans the next day, we realized that as it was the dry season, most of the animals, including lions, were far away in the Okavango Delta. We needed to go into the swamps to find them.

That meant a drive of about one hundred miles back in the direction from which we had come. "West toward the delta," we chanted, taking off again. The nameless sandy red track led to a crude causeway of stones and Mopani timbers. The baTawana settlements crowding around the small administrative settlement of Maun were the first sign we had reached our destination. It was the gateway to the Okavango swamps and home to hunters and poachers, among others.

Trees grew around the water and the bridge. A large number of round huts suddenly appeared before us like stemless mushrooms after a rain. They had no protective stockade around them. These belonged to poorer people, those who worked as servants and others with no cattle to protect.

With the exception of the major sand road we came in on, there were just odd tracks weaving in the red sands around huts. All was overprinted and trampled bare; only sporadic acacia trees grew.

Near the village center, a half mile or less of tar had been recently laid over the track—a sign of progress. Before this, only a four-wheel-drive vehicle was able to pass through. Kicked-up sand would clutch the wheels of other vehicles until they sank spinning. Time tested ox- and donkey-drawn sledges did pass, churning the sand like making milk into butter. Riding on a smooth road felt weird after so many miles of washboard sand tracks. The section led to an assortment of European-style homes, trading stores, and a hotel. These one-story buildings, arranged side by side, lay inside a compound facing an inner courtyard. We drove in and pulled up in front of the hotel. As our engine sighed, so did we.

"There's a bar here," Patrick said, perhaps forgetting he had mentioned it a half-dozen times already that day.

We walked inside, rather wide-eyed, and stood in silence in front of a long line of wooden barstools.

"My ass hurts," Marcus said.

It was midday. There was only one other customer, a gray-haired old-timer dressed in khakis, sitting firmly on his stool and sipping a beer. His wide-brimmed, sweat-stained hat lay next to him on the wooden bar.

The entrance walls were decorated with crowded configurations of horns, skins, traps, and other wildlife mementos. Bottles of labeled and unlabeled brew rested behind the bar.

A cold beer, homemade and gassy, was handed to each of us in response to Patrick's anxious gesture of nodding the full upper half of his body at a tap. Marcus added shots of whiskey as he gulped the beer. I struck a match and lit my pipe, and the old-timer sniffed with a show of appreciation. We caught each other's eyes. He had a tough weather-beaten face and carried a small knife on his belt, the size for gutting and skinning. He was a white hunter, maybe a poacher, I guessed. I asked if he'd like some tobacco.

"*Dumela*," he replied, using the local word for greetings. "Sure, thank you."

Drawing on my pipe, I sat down next to him and passed my pouch over. As he struck a match, he said, "The River Bushmen believe that the ostrich first brought fire to the world. The bird kept it hidden under its wings so no one would steal it. Man tricked the bird—I don't recall how, maybe the myth didn't say—and found out the secret, but even today, the ostrich will not use its wings to fly for fear that other animals may steal the secret, too."

Yeah, he was a white hunter, a stalking guide for those hunters who could afford it—a fascinating, dying breed in Africa. Animal stories and sage-like stories were part of their well-known campfire character. I pictured him with heavy-caliber cartridges in loops across his chest.

Patrick joined us. "I once heard that the Ila people of Zambia believe that fire was brought by a mason wasp." He sat down and leaned around me to look at the hunter.

"Are those the ones that are yellow and make mud nests everywhere?" I asked.

"Especially on fireplaces, where it's warm," the hunter said.

"So what's the story, Patrick?" Marcus asked, taking a step closer as we all took a sip of beer. I ordered another for the hunter, hoping he would stay, although I needn't have worried about that. At that point, we introduced ourselves. His grip was firm, and I felt the calluses on his hand, including his index trigger finger.

Patrick settled further into his barstool, holding his beer. I guessed this was a story he had heard in an Anglo exploration camp. Sometimes outdoor people are judged by the stories they know.

The white hunter sat, sipping his drink. If he knew the story, he didn't say so.

"In the beginning, there was no fire on earth. All the insects and birds gathered and decided that to keep warm, they would ask God if he had fire. The mason wasp said he would find God and ask him but wanted somebody to help him. Three volunteered—the fish eagle, vulture, and crow.

"All flew high in the sky for many days. Finally, some bones from the vulture fell to the ground. Next the bones of the fish eagle fell, and a day later the small bones of the crow fell. For a month or more the mason wasp carried bravely on without his friends. He rested when he could on the clouds but couldn't reach the top of the sky. God heard about this, however, and decided to come down and talk to him."

Patrick stopped, gulped his beer, and continued as we all leaned on the bar, listening.

"The mason wasp explained that he was trying to reach the Chief of All to ask for some fire, but he wasn't able to fly any higher. He sadly said his friends had disappeared behind him. God felt sorry about all this and said he would give humans fire. He also said that since only the mason wasp had reached him, the wasp, as a leader, would be allowed to live all over Africa, and whenever he wished, he could build his nest near a fireplace to keep his children warm. To this day, the mason wasp does this, as God gifted him."

We sat silent for a moment. African stories were like drawing a figure from a group of stars. You had to want to see it.

"Do many Bushmen still exist in the swamp?" I asked the hunter. "Are they still nomadic?"

"When I first came here fifty years ago, there were many, many more than now. Most of them have intermarried with the black people who arrived here after them. A few bands that I know of live to the northwest in the Tsodilo Hills, also called the Slippery Hills. It's an island of rock that juts out of its sandy grave. The hills are a sacred place—a place where the Bushmen believe supernatural things happen. They think God created all animals there. It's a very dry area, but springs and seep-like watering holes pop up from nowhere. The Bushmen honor that phenomenon religiously and seem to have settled there now.

"I think," he continued, "they would have great difficulty surviving if they had to migrate as they used to. They are a dying race, like the white hunters, and unable to adjust. This final stage comes after they have survived some four thousand years of being chased all over southern Africa by man, leopards, and lions and enduring a terrible climate."

"Did you hunt?" I finally asked.

"For some fifty years I shot big game over much of southern Africa. I guided trophy hunters from all over the world and killed for ivory, too. It hasn't been easy to hunt, though. In the late 1950s, foot-and-mouth disease broke out repeatedly, halting the cattle industry. In an effort to control the disease, veterinary surgeons had wire fences stretched over enormous distances. Water sources like the Boteti River were cut off. Masses of animals began to pile up before the fence, unable to reach the water. The timeless migration routes were blocked. The days of free-roaming wildlife in that area were finished. It wasn't the white hunter but the damned cow who saw to that.

"Cattle overburdened the grasses with regular feeding. Wildlife feed only a few times, if that, in a given area in one year, and then wander. In some areas the grasses have been eaten off and taken over by worthless weeds and thorn scrub."

Another environmental disaster caused by humans, I realized.

"Still, enormous tracts of land remain unaffected," he continued, "including large portions of the delta itself, which the government has made into game reserves."

"The Moremi Game Reserve[3] in the delta," Patrick chimed in.

---

3 Known then as the Moremi Wildlife Reserve and Chobe National Park

"Yes"—the hunter nodded—"and the Makgadikgadi Pans Game Reserve, Nxai Pan National Park, and, of course, to the south, the Central Kalahari Game Reserve. The old days are gone when I could roam at will. None of this stops the poachers, however. But I've had a good life. I've a wife or two, children here and there, and I've made enough money to live without hunting more. I also enjoy the odd drink." Patrick understood and motioned for the old-timer's glass to be filled.

"I enjoy the silence here, the lack of problems," he continued. "There are no thieves, and people share."

Marcus finished his beer and signaled for another round to include whiskey. "It sounds as if the animals go from one crisis to another."

"Only in rare moments can they can relax and feel safe with a full belly and ample water. If it weren't for the danger and feel of hunger, though, they wouldn't stay alert and alive."

Marcus downed his shot; he was getting quite plastered.

He wiped his red beard with the back on his hand. "If an Australian bulldog ant is cut in two, a battle immediately ensues between the tail and head. A cockroach can be without a head for several days, and then it starves to death."

"An odd contribution, Marcus. What did you do on that farm? Chop animals up?" Patrick asked, leaning over the bar to see him as the rest of us chuckled.

"Do you think Livingstone had much effect on the natives?" I asked before Marcus could answer.

The old man shook his head. "But he opened up the route here for the traders and hunters who came and shot the animals. That certainly influenced the natives. They soon realized that ivory was worth beads, cows, and such objects they wanted. Livingstone's main purpose was, as a missionary, to convert the natives to Christianity. He was no cultural leader, and I doubt he converted many people. Most here will tell you, including the missionaries themselves, that Victorian virtues such as thrift, one wife, and honesty were all—logically—avoided by the natives. Add this to the concept of resurrection and things like equality, and the black man was completely baffled and has basically remained so. Can you imagine an African chief being convinced that those he killed in battle might rise up again?"

He struck a match and relit his pipe. "They probably thought the white man's religion was just a trick to get them to give up their wives. Influence within the tribe was a way to survive, and if you had many wives and dependents, you had a better chance of making it. Also, many wives could grow more food, which meant more food for festivities. Then there were the associated gains. You could marry your children off to other headmen and possibly win their support in the process. As to being thrifty, they had to share some things, like grain and cows. It was expected. Equality was another problem. They didn't like the idea of Livingstone trying to convince the average person he was as good as his chief. For all that, he wasn't harmed by anyone. He was more a thorn in the side of the natives, and they were used to thorns. On the other hand, Stanley, looking for Livingstone, shot natives like animals."

The bottom line is that the white man measured the natives on the basis of European standards and laws. They had no interest in viewing them as they were with their own culture. Livingstone died at age sixty while searching for the source of the Nile River. His right-hand companion, Dr. John Kirk, came to the conclusion Livingstone was mad.

I was fascinated that the local people apparently hadn't changed over the years. "Have they learned anything from the white man?" I asked.

"Sure," he said, his eyes twinkling. "Many of them travel south to work in the gold mines around Johannesburg. They come back with radios connected to car batteries, which they carry under their arms."

George and Patrick digging for diamonds near Bushman's grave

George and Land Rover in Namaqualand

George's field area in Namaqualand, South Africa

Namaqualand

Home of a native working on Afrikanier farm

The Bushman's grave, South West Africa

The swing at the Oasis

Window in the Oasis home

Blue lace agate, South West Africa

Dark metabasite rock with sledgehammer, Namaqualand

Tombstone tribute to a fishing boat that went down with all hands, off the South Arican west coast near Dog Rock Bay

Metabasite Rock Namaqualand missing piece I took for sample

George in Cape Town

A Namaqualand quiver tree

A weaver bird nest on the shore of the Orange River

George at Fish River Canyon

George swimming in Fish River Canyon

The gemsbok the morning after it died at a watering hole

Gemsbok, South West Africa

Sand dunes, South West Africa

George and Patrick digging for diamonds, Forbidden Area

Keetmanshoop ghost town, South West Africa

George clowning around on former stage in
Keetmanshoop ghost town, South West Africa

Keetmanshoop ghost town, South West Africa, Manager's former home

George at old swimming pool, Keetmanshoop ghost town

George bathes with elephant in Okavango Swamps, Botswana

Native children play in in crocodile-laden Okavango Delta

George bathing in the Okavango swamps, Botswana

Convoy in Rhodesia, our Land Rover just passed (left as we were too slow) by rear guard of convoy

Author at Victoria Falls, Rhodesia

Exfoliation Domes, Namaqualand, same kind of rock Backside Bluff was formed from

Namaqualand showing exfoliation domes

Recent shipwreck (1974) off coast of South Africa

Xhosa hilltop home, Transkei

Xhosa hut in the Transkei, South Africa

George receives Ph.D. from the University of Natal, South Africa

Chapter 16

Into the Largest Delta in the World

After taking a final pee at the bar, we stumbled into the sun outside, realizing we were about to penetrate one of the largest swamps in the world while somewhat intoxicated. Nevertheless, Patrick began filling our Land Rover with gas while Marcus began contaminating the immediate air around him with his own gas. His farts sounded as though air was burping from a truck tire.

The vehicle's white finish was covered with desert dust and our handprints as we staggered around, leaning on it. It looked like bizarre wallpaper. As Patrick poured water in the radiator then filled our water containers, I caught sight of Marcus gagging a few steps away.

Climbing awkwardly into the Land Rover, Patrick started the engine, mumbling, "No problem with drunk driving here."

With Marcus facedown on the backseat, we followed the sand track to the border of the wetlands. Next to the wood and stone bridge we had passed before, a group of children were playing in the clear water. Their skin was very black and their teeth very white as they laughed and paddled amid patches of green reeds.

Water lilies, with trailing stems below the surface angled in one direction, demonstrated a slow and gentle underwater current pushing at them. They open their petals in the morning as the sun rises and close them when darkness falls.

We knew crocodiles were there. The children seemed to have no fear, although we knew from time to time one of them had been lost to a crocodile. It was a bit colder now, and the reptiles had a limited appetite when their metabolism slowed. The people lived with the threat, perhaps like Western people living with the possibility of a car accident.

A shapeless old woman appeared, collecting Mopani worms from the Mopani trees next to our track. We stopped to watch; she peered into the backseat apparently curious about Marcus. A plug of wood protruded through each of her ear lobes. The worms looked like finger-size caterpillars, and she had a small basket of them. She pinched one at the tail and squeezed out its yellow flesh as if it were toothpaste. The insect's green innards slimed out on her two poorly manicured fingers; she held the carcass before her open jaws for detailed inspection.

Just then Marcus sat up—she was blocking the sunlight—and looked directly into her mouth as she bit one and chewed. "*Agggggg*," he moaned, then yanked on the door handle and puked outside.

Patrick immediately drove away as the odor rose. "Can't very well throw him out of his own Land Rover," he grumbled. "Better to pull over and just go to sleep."

And so we did.

* * *

Morning came, and our eagerness returned as we gnawed on roasted corncobs purchased at the bar. Then we drove directly into the swamp.

It was like entering another world, with trees brushing against the Rover, the demanding sun reaching us only in patches now. Patrick and I lay on a mattress on the roof rack of the Land Rover, surveying the surroundings, as Marcus, still not clear-headed but wanting to drive, moved forward slowly over the rough track.

Then, as if an invisible forest curtain suddenly rose, they came. Like World War II kamikazes taking aim at ships, they descended on us in swarms: silent, light-footed tsetse flies. Each was the size of a corn kernel with wings. They had engulfed us unawares, until their painful red-hot needles stabbed into bare flesh. A black cloud of unrivaled passion, they attached to us like frenzied lovers. Exposed hands and faces were not enough for them, we soon realized—they even bit through protective clothing.

"Give it stick, Marcus!" Patrick screamed, hanging downward, looking at our driver. "They're tsetse flies!"

We swatted at all angles, trying to protect ourselves.

"Hell!" Marcus shouted as he pushed the gas pedal. The draft quickened, and the flies were unable to land—though some had already dug into our shirts and hair.

"One is down my back!" Patrick cried as he twisted, trying to reach it.

I hit him hard with my palm as we bounced along.

"Shit, man!" he called, losing his balance and falling toward the edge of the roof. His arm dangled next to Marcus's face.

Marcus hit the brakes. "What the hell is happening? Is a monkey on the roof?"

Patrick and I lurched forward as the Rover halted and the flies landed again.

"There's a monkey *driving*, for Chrissake!" Patrick yelled, still hanging over the side and holding on while being bitten by the flies. "Don't stop, man, don't stop!"

Marcus, now also being bitten, hit the gas again, causing Patrick and me to jerk backward, and we were now barely able to hang on to the metal roof rack.

"Keep going! Keep going!" Patrick yelled in agony, his face flat against the mattress. We sped forward just fast enough to keep the flies off and still negotiate the track.

A mile or so later and rather suddenly, the flies stopped, as did Marcus. "Jesus, what was that about?" I said weakly. I looked at Patrick. We were lying flat and still holding on, for fear Marcus would race forward again.

"If you mean Marcus, he seems to like the brake and gas pedals," Patrick said weakly. "Otherwise, we passed through a fly belt. They spray the swamps for the fly but leave a protective perimeter." He opened his eyes and slowly raised his head.

Marcus got out and looked up at us; the red spots on his face meant he'd also been bitten. "Little savages," he growled, then he gagged viciously several times.

"What's wrong with you?" I asked. "Are you still sick?"

"One flew into my mouth when I yelled up to you. It's in the back of my throat." He bent over, hacking.

I jumped down and started slapping Marcus on his back to help him clear his throat. The flies weren't dangerous, I told him, unless they carried parasites that cause the drowsy sickness. The parasites come with the blood they ingest from an infected host, and then they pass that on to whatever they bite next.

The chances that we'd become infected were slim. There had always been controversy concerning spraying. If there were no flies, the local people would bring in cattle, which would eat everything and turn the land into a wasteland. Fences, buildings, and cultivated land were next. The tsetse fly helps Africa remain Africa. The fly belt is like a cattle fence.

"It's as if there's a peanut in my throat," Marcus grumbled, climbing on the roof.

"I assume it's dead?" Patrick asked.

"What do you think? It's shimmying up and down?"

A bit later, as Patrick slowly drove along the ribbon-thin grass-covered track, I peered into a stand of tall Mopani trees we passed. There, with his trunk moseying up and down a tree, stood a huge elephant.

"Look at that, Marcus!" I grabbed his sleeve and pointed. Neither of us had ever seen a wild elephant before. I tapped on the roof, and we rolled to a stop to watch the massive animal in silent awe. He was a marvelous piece of equipment, pulling branches down to feed. He pushed his massive head against a tree and uprooted it.

"It's as though he was nudging a stuck door open," Marcus replied.

Able to reach the tree's branches now, the elephant began stripping them of leaves and bark using his all-muscle trunk as if it were a backhoe. He stuffed it all in his mouth and then turned to watch us. His ears stayed back, flat against his side, suggesting, I presumed, that he wasn't overly concerned. He defecated without pause, and vapors rose from the pyramid pile—a monument to the deceased tree.

I wondered how many seeds were in the pile and if a new tree would grow from it. That was one way the forest replaced the trees that the beast knocked over.

"He seems relaxed," Marcus suggested, observing the steaming pile.

"He has learned to eat and shit at the same time, unlike most of us," Patrick whispered up to us.

"It's remarkable there's no fence around him," I said quietly. "He is free, absolutely free."

"An elephant his size has to eat five hundred pounds of greenery a day," Patrick whispered up to us. "He doesn't digest all of it. That's why he's crapping so much."

"Why don't you leave him a few of your old potatoes?" Marcus suggested. "They might even taste okay to him after a meal of bark and leaves."

"Why not leave him your underpants, Marcus? There's probably something green growing in them."

Patrick started the engine. It sounded curious in the forest world, like a clashing of cymbals in church on Sunday. The elephant's ears came forward a bit as we crept away. "I hope he doesn't chase us," I said to Marcus. "They can run about twenty, twenty-five miles per hour, and with his long trunk and us up here, we would be like low-hanging fruit."

An hour later, as we turned a corner, we saw the large backside of a massive, armor-plated black rhino ambling down our track, almost prancing with power. Patrick immediately cut the engine, and we stayed still. The black rhino, we all knew, was an inquisitive, irritable, dumb, and excitable animal that will charge at the slightest provocation. Some say that an earful of ticks will drive the animal mad; and it won't run from an enraged elephant as all other animals do.

We lay quietly, watching, wondering what he would do. He solved this by shuffling partially around and looking in our direction. Peering over his thick horn, turned upward like an erection, his accentuated hard barrel of a belly swayed to a stop. He couldn't see more than fifteen or so paces ahead of him. The breeze was in our face.

"He's blinking and thinking, like an old boxer that's been hit too many times," Marcus whispered.

"He can hear us," Patrick added.

I wondered, lying on the roof rack, if my friend had a plan should the great rocklike hulk turn and charge us. The rhino was like a small locomotive, and we were a carriage on his tracks. Nothing would turn him once his whistle blew.

He could run some thirty, thirty-five miles per hour at top momentum, I figured. That might be our maximum rate, too, on this pitted and twisting track . . . and we would still have to turn the Land Rover around before we could try to get away. Not enough time? Patrick wouldn't play chicken with the monster, would he, and try to make the beast back down? I looked at Marcus and he at me.

"My Land Rover," he said with wide eyes.

Oh shit, Marcus was thinking the same thing—rhinos do not back down. No matter how tightly we hung on, we would rocket off the top if the rhino hit us.

Five minutes passed with no movement except the rhino turning his head slowly from side to side as his little eyes stared at the car, wondering . . . Then he slowly turned, lowered his head, defecated digested bark and twigs, and moved off into the trees.

"I think he forgot about us," I said.

"He has a small brain," Patrick yelled up to us.

"What did you plan on doing if he came for us?" Marcus called down, partially hanging over the side. "Beep the horn over and over and flash the headlights?"

"Probably just turn on the windshield wipers and wait until you started gagging again. That would turn anything," Patrick, sticking his head out his window, yelled back.

We continued. Sudden movements flashed in the forest as hoofed animals and others foraged.

In the 1970s, there were some crude, ill-defined areas in the forest for travelers who wished to camp. We settled down in one rough clearing for a peaceful evening of frying steaks.

Our campfire was tiny compared to the one we had on the Makgadikgadi Pans. The shadowy forest canopy made it seem even less significant as it hung black and thick, obscuring starlight. The rising smoke appeared ghostlike when wind rustled the trees. Light melted away a mere step from the fire.

There are different degrees of darkness. The forest that night was in the *rat-black* category, which means particularly dark. It's the kind of darkness that exposes all men's behavior. If a man is afraid, he will show it.

The evening silence was broken by sounds that rose and fell, creating an irregular forest symphony. Unseen phantoms rustled leaves and broke branches, growling, moaning, breathing, and betraying themselves under the blanket of night. It could be a mouse scurrying to find his hole, an elephant trumpeting the sunset, a monkey shaking a branch, a jackal whining, a lion roaring, a hyena crying. Even a tiny beetle struggling to pass through dry leaves made an astonishing noise. In the background is the constant chorus of crickets and thousands of insects, each with their diverse songs of life. And when the moon is full and sleep confuses them, birds add their own contributions. It was an African night in a swamp, a beautiful and frightening time, depending on who you are and what fate will provide you.

Nocturnal fireflies dancing in the warm air began to disappear. Dawn came, pushing away the darkness. It was announced by the high, clear duets of a few impetuous birds, apparently hysterically happy to have survived the night. The thought of having a breakfast of leftover steak, eggs, coffee, and Patrick's potatoes lured us out of our sleeping bags. After eating, we lay on our backs. Fingerlike gray clouds shook hands as they met this quiet morning. They would soon say goodbye as the warm sun drove them apart. We lazily crawled back into our bags and fell asleep.

I was only partially awake when I heard Patrick quietly say, "Don't move."

"Oh shit." I opened my eyes and tried to see under the rim of my hat, which was partially over my face. "Oh please, no snake."

"What? *What?*" Marcus whispered; he was facing the other way and couldn't see.

"Giraffes to the right," Patrick whispered.

I heaved a deep breath. "What a relief." I turned slowly to look and then gasped. "Will you look at that!"

Only fifty feet from us two very tall giraffes had moved into the clearing, twisting their patchwork necks and eating favored leaves off the upper branches of acacia trees, which offered two seeds in each green-jacket pod.

"I think adults giraffes normally don't travel together. Maybe they're a pair, and the female's in heat?" Patrick said quietly. "Males taste the female's urine to see if they are fertile. He could be waiting for an opportunity."

"Unbelievable the things you know," Marcus replied.

The upper and lower portion of the giraffes' frames did not seem to be synchronized, which resulted in an ungainly walk. I noticed then that both right legs moved forward then both left. They had to shift their bodies like a human with wooden legs, while their swaying necks and heads provided balance. Still, they presented dignity as they scrutinized each tree.

"The thorns of the acacia don't seem to bother them," Marcus said. "They are just shoveling those leaves in . . and they are also coming closer."

"Their tongue is special—all muscle and long, maybe a foot and a half, and flexible," Patrick said. "It curls around and pulls off what the beast wants. Look, one is licking its ear. That's a hell of a tongue and it's purplish-black! I wonder if they stick it into a female giraffe's ear?"

"That's an odd thing to wonder."

"It has to do with sex, Marcus. Actually, the elongated ears do look like a vagina," Patrick continued, squinting in the sun and not paying attention to Marcus.

Marcus looked at Patrick in disgust. "Do the Irish really think like this?"

"Amazing how strong their legs are," I said to change the subject.

"The giraffes don't have claws or upper front teeth," Patrick said. "They use their legs, thick with muscle, for protection. They kick like a man in a street fight, only backward."

We later learned an adult giraffe has few enemies and can break a lion's spine or shatter its skull with a good kick. I suppose being struck by a giraffe's leg was like getting swung at and hit by a very long, heavy pole. Maybe that had something to do with their walking style.

At that moment one of the giraffes began peeing. The urine had a long way to go to reach the ground. We watched it flow as if from a garden hose and then spatter, when the two-story animal stepped toward us.

"Oh hell . . . it's going to douse us!" Marcus gasped. "The other one's following."

No one commented further; we were scrambling, still half in our sleeping bags, toward the Land Rover.

Chapter 17

From Dugout Canoes to Military Convoy

To enter the swamp, we needed three dugout canoes, or *makoros*, carved from large wetland trees. The Hambukushu—one of the five ethnic groups in the delta—were our "motors." Standing in the rear of each dugout—loosely dressed, barefooted, thin and wiry with protruding knots of muscles and drooping hats—the Hambukushu men poled us forward. Our entry point was lost a minute later behind a shifting fog of greenery.

Looking at my map, I traced the broad blue line of the Okavango River, which feeds the swamp, back over five hundred twisting miles to the faraway mountains of Angola. For some six months after the annual rainy season, the river flows incessantly south over the dry Kalahari. *What,* I wondered, *could be more determined than a young mountain river?* When it finally reaches the flat delta, it fans out into channels. But from there it has nowhere else to go. In the end, the Kalahari sands simply absorb it as if it never was.

There is some geological mystery to this, however, as not all of it sinks into the hungry sands. The new water, buoyed by deeper prehistoric fossil water, floats outward, doubling the size of the swamp.

It engulfs the salt pans, surging as far as it can before finally giving up and sinking or evaporating.

In addition, the channels change every year as the land imperceptibly tilts from deep tectonic movements underground and the waterways become choked by the growth of new papyrus trees. One day there are defined water roads into the swamps, and the next day the marauding flood obstructs all, creating a maze.

We poled on into a stream channel, staying together. The shore was somewhere between the green water sedge—tufted marsh plants—and reed beds and the enamel-green land-grass beds. It all blended like a seamless puzzle.

Painted spiders configured cobwebs that dangled with grace from tall reeds, their center points quivering with the faint breeze. The morning sun captured their misty walls silvered over with diamond dew, readying them for transformation into deadly invisible traps to capture a multitude of flying insects that inhabited the swamp community. There were hundreds of webs in every thicket, some old, some new, and some in the process of completion or repair. Invariably, the odd white stork or a similar bird would emerge from the reeds, striding along on its thin twig-like legs, fishing. Sometimes the birds would jar a spider's sticky web and eat its creator in one fast movement.

My dugout leaked a bit and had been patched with flattened tin cans. One said "*Coca-Cola.*" I sat on the bottom, feeling the water rise; it was crystal clear. Several yards down, deep-green vegetation wound its mesmerizing way upward. All were willing slaves to the currents that transported nutrients and life.

Thousands of wattled starlings passed overhead in a huge flock, gulping insects as they flew, a dark thunderstorm blocking out the sun. Suddenly, hawks from above them dived down, tearing through the now-trembling cloud-like falling arrows. The starlings bolted, blown apart by terror. A moment later, the blue sky reappeared.

From time to time vultures sat, each head hanging on a looped neck, in the dead trees that dotted some areas of the swamp. A macabre sight: the undertakers dressed in black and white sitting in skeletal, barkless trees. Their bodies remained immobile while their heads turned, watching us pole past them. I could feel their disappointment.

"You can't eat them. They stink horribly," Patrick said.

Marcus and I looked at each other, not saying anything.

"Look! We're going over a hippo channel." Shading my eyes to see better, I stared downward. The bottom grasses were trampled where the huge beasts walked, apparently preferring to use their channels repeatedly rather than forge new paths. It was their road system through huge beds of uninterrupted reed vegetation. Staring downward, Patrick said, "Hippos kill more people than any wild animal in Africa." Unconsciously, he moved a bit in his makoro, causing it to rock. The pole man immediately changed his balance in the boat and held his pole against the swamp bottom to steady it.

"When we were in that little tree you said buffalo were the most dangerous . . "

"It depends on the situation."

Sometime later, our channel opened up into an expanse of water. Around the perimeter floated masses of green *Trapa natans* plants with their tiny air bladders. They dulled the harsh reflection of the sun off the water. As we poled away from the plants, the reflection intensified, hurting our eyes and obscuring our vision, even with the shade provided by our broad-brimmed hats.

Shielding my eyes with my hand, I peered at a white sandbank ahead, its water-polished quartz grains sparkling like thousands of tiny mirrors. "Good Lord," I said, staring.

"Nile crocodiles," Patrick said.

The bank was covered with crocodiles lying in the sun. Some swam leisurely, making small circles in the water with their stubby webbed feet. Due to their good night vision, they normally hunt in darkness.

"Maybe crocodiles are the most dangerous," Marcus suggested.

"As I said, depends on where you are," Patrick replied.

"And look, those are hippos floating over there!" I pointed near some papyrus islands. "And here we are on a sinking log. Shit, I hope we're not going to paddle any closer."

My pole man turned around to look at the native in his canoe. We stopped. While the guides could not understand what we were saying, they apparently detected our distress and agreed with us.

We were close enough to see tickbirds pecking on the backs of partially submerged hippos and sitting on the open mouths of the crocs that tolerated them as they picked away at the rot between the reptiles' teeth. As they selected their morsels, they sang their tickbird song.

"A fine example of symbiosis," Marcus said. "I wonder how many birds were killed before nature got that plan to work."

All I could see of the hippos were their ears, eyes, and nostrils, placed high on the skull to allow the bulk of the animal to be submerged, like a submarine with its periscope up. In places, in groups of ten or more, they were obviously standing on the bottom. Their skin does not have sweat glands, so to prevent dehydration they stay in the water submerged or partially submerged in daylight. Some grunted. Others rose and sank. Still others slept—they had the ability to do that underwater and rise automatically to breathe, only to sink again.

"They are adults, the little ones stay away as the older ones will chomp them up," Patrick said.

"Kind of crude," Marcus added.

"Do you know hippos pee backward in the water, leaving a scent until it dissipates?" Patrick continued. "They also spin their curly tails, throwing their shit in different directions as it comes out. It sinks to the bottom like a marker."

Just then a fish eagle screamed overhead. Instantly, all eyes were on it. Even the pole men watched with wonder. With a wingspan of some seven feet, a white headdress and tail, and powerful black wings that sliced the sky, it commanded attention. This one had apparently been watching the water for movement. It swooped down with its large talons opened like hands with curled fingers and razor-sharp fingernails spread wide. The selected fish was huge and so heavy the bird could not raise it more than a few feet off the water. Rather than let it go, the bird dropped into the water and tightened its grip. The fish flapped madly. Water sprayed in all directions.

The strange battle raged on—the huge sky bird yelping distinctively and paddling to shore with its powerful wings while clutching the struggling fish that had only moments left to live.

Floating silently for several minutes after the skirmish finished, the makoros maneuvered around long, thin islands that were curved like the meandering river they were once part of. When we reached the shore, we carefully exited our rocking tree trunks and gave our pole men a pula or two (worth about a dollar then). Following the track several miles back into the swamp, we found a small clearing that surrounded a thick growth of tall leadwood and shorter sausage trees. Cork brush dotted the areas between them with other bushy foliage we couldn't name. There seemed endless varieties of flora growing within the sprawling floodwater delta. I wondered if anyone could recognize them all.

Hunger pains took priority over such thoughts; it was time to eat. Frying the last of our meat and baking potatoes, we talked as young men do, knowing that life is before them and around them. The setting sun presented itself as a raging red fire; it enthralled us to the point of silence. The next day or so was spent at one of many islands, watching animals come and go and clouds drift across the sky.

* * *

It was time then to consider our next destination: Rhodesia. We had last discussed that as we planned this trip while eating chicken in my kitchen some three weeks ago. Now that we were near its doorstep, should we enter a country in the middle of a savage civil war?

On one side in the struggle were the black guerrilla fighters of Joshua Nkomo's Zimbabwe African People's Union (ZAPU), the majority of whom were local Ndebele people. The opposing force was Robert Mugabe's Zimbabwe African National Union (ZANU), mostly composed of local Shona people. As two separate armies, the ZAPU and ZANU fought the white-controlled Rhodesian government of Ian Smith. All three groups wanted to govern the country and pursued the war with increasing ferocity. The Rhodesian Broadcasting Corporation reported that ZAPU and ZANU were shooting each other, setting off mines, laying ambushes, cutting off heads and arms with machetes, and burning families to death.

The war was at its peak. In less than a year from now, Robert Mugabe would come to power and the country would be renamed Zimbabwe. Some regarded him as the worst dictator in the world. Over the next thirty years, he would demonstrate that no one can kill a black man faster than another black man. Most of the surviving whites were driven from the country. The resulting poverty and deprivation were conveniently blamed on the Western world. The larger the enemy, the more the pity and glory one could claim.

"Well, shall we continue?" Patrick asked. We knew that entering Rhodesia in our old Land Rover would be dangerous.

"We may never get another chance," I answered, palms up. "I've heard it's a beautiful country. Who knows what will happen to it?"

"Okay, let's go," Patrick said, shrugging and, I suspected, thinking about the white hunter's gold nugget from the Rhodesian stream I told him about. He put the Rover in gear and off we went.

That is a magnificent thing about youth; it can decide quickly—before the logic and fear of consequences that come with age can intervene.

The road from Maun, Botswana, to Rhodesia went north-north-east into the Chobe National Park (CNP), or east to Nata and then north or south. The CNP road required a four-wheel-drive vehicle even in the dry season. Our Land Rover was too old for that rough terrain, so we chose the tarred Nata road and then the drive north.

At the end of the Nata road leading north we drove first to the town of Kasane and then through the settlement of Kazungula, which our map showed to be on the Rhodesian border. Most of the way to Kasane no one spoke. We were all thinking—worrying, no doubt—about our decision to enter Rhodesia. Patrick did advise us that he thought Richard Burton and Elizabeth Taylor wedded a second time there in 1975. "Outside of the district commissioner who married them, there were no other guests."

Marcus and I looked at each other. "Hard to believe they drove in the way we did," Marcus added.

Patrick clarified. "I think they used a private plane."

"Look!" I pointed in awe as we drove. "The Zambezi River . . . another of the rivers I dreamed about as a boy. Hell, it's big! And to think it began as a trickle far to the north, close to the borders of Zambia, Angola, and the Congo!"

A ferryboat took us and our Land Rover to the Rhodesian side of the river. From there we drove forty miles or so, where we recrossed the Zambezi on the Victoria Falls Bridge and entered the small town of Victoria Falls itself. A female tourist had been killed by shots coming from across the border a month or so before. The residents carried on with their lives as best they could. Perhaps it was like the natives living with crocodiles. Maybe you just *have* to get used to such things.

Having errands to run, the three of us went our separate ways for the remainder of the day. I set out for the falls, walking over the flat plateau that surrounded it, hearing its muted rumble from far away and following a disused path through trees, weeds, and bushes.

The rumble steadily became a roar until suddenly it burst into view. I gasped—Victoria Falls! More than 350 feet of falling water spewing mist like steam from boiling water. It was twice the height of Niagara Falls, the *biggest* curtain of falling water in the world. More than a thousand tons of water per second, on average, hurtling forward. Some said a meteorite had created it. Well, it deserved some kind of extraterrestrial association, I decided. The locals named it "the smoke that thunders."

I marveled as water raced to the edge of the falls only to suddenly plunge over it. Seconds later it smashed into the distant bottom, exploding upward as if in a wild rage. Falling again it regrouped and charged forward with groping white claws, slashing and twisting around obstacles. Surging with huge force it smashed and crashed its way out of sight. Then, it was the Zambezi River once again—a massive, calm, moving part of the earth, free and wild, a gift of life from which history and folklore flowed, a measure of God's finger as his hand carved the wilds of Africa. On it would flow for hundreds of miles, ending in the Indian Ocean, where its mystery and excitement die with a whisper, just as all life dies and just as it all begins.

The falls mark an edge of the great central plateau of Africa. It was the border of darkest Africa. It was a place where something began or ended; there was no middle.

As I stood staring, I noticed a statue in the thicket to my left. It was of David Livingstone. A plaque read he had first seen the falls from the point where I stood. I looked at him as he stared with the deep eyes of an owl; a man of unbelievable if not appalling determination, who lived mostly in his own world as he pursued his goals, even going somewhat mad. Thinking of myself, I wondered if following dreams with such conviction was good. Maybe it was better to know when to stop?

Our plan was to travel south to Bulawayo on the two-lane macadamized road. Due to guerrilla activity, the only way to do that was by joining a convoy. The idea was that a fast-moving target was harder to hit than a slow one.

So we joined one. In our convoy, the lead and last vehicles were open-back pickup trucks. A helmeted and sweating main gunner stood in the rear of each truck, clutching the pistol grips of a pole-mounted, chest-high M60 machine gun. Unfortunately, our Land Rover, placed last in line in front of the armed pickup, could not maintain the convoy speed, so we held the others back.

"Give it stick, Marcus!" Patrick said, as our engine roared to full capacity and the trailing gunners waved at us to hurry. So concerned were they that at one point they pulled up next to us. The Land Rover rattled. Baling wire that held on a fender was coming loose. It was like being in an old washing machine. The tension increased proportionally.

"Lord Jay-sus help us now!" Patrick yelled. Marcus said his leg muscles hurt from trying to pin the gas pedal flat to the floorboard.

There was nothing to do but watch uncomfortably with fisted hands as we roared forward, the main gunner training his weapon on successive clusters of rocks and trees dotting the adjacent hillsides. Death could be behind any of those hiding places. His partner, also clad in brown camouflage fatigues, crushed the butt of his cigarette to powder between his thumb and forefinger. He reduced his profile by kneeling at the gunner's side. Poking what looked like a lightweight semiautomatic over the side of the pickup, he watched in the distance for anyone suspicious.

These men were part of the citizen army, essentially composed of any adult male available and willing to take their turns (six months army, six months at home) to help keep the country from falling into lawlessness. I wondered how good a shot they were.

Eventually, with the caravan speeding ahead, we neared Bulawayo. The armed pickup again pulled alongside us, but this time it kept going—leaving us behind like horse manure in a circus parade. Bravely, we chugged and rattled our way into the city.

Where to go next? We all were thinking the same thing: We would look for gold. The rest of life, our duties and problems, could wait.

At the same time, my desire to continue my studies burned inside me. It just wouldn't let go, wouldn't let me say the hell with it.

Chapter 18

Gold in a Gully, Vultures in a Tree

Over a beer, I told Patrick and Marcus again how the old Rhodesian described the southern route from Bulawayo to the "gold stream" he had found some fifty years earlier.

"Some of the richest gold mines in the world are to the south, around Jo'berg," Patrick said, "and gold workings have been found in Rhodesia as far back as the 1860s. I used to study them during my gold-prospecting days in Botswana."

I thought back to the afternoon I'd sat with the old prospector, in the shade of the hotel porch in Springbok. "The gold is near a distinctive granite exfoliation dome," he'd told me. I listened intently as he went on to describe what the structure looked like. The old hunter finished by saying, "I don't want to take its location to my grave—I'd rather share it with others." Such stories of lost treasure told by men facing the end of their lives are not unusual in untamed Africa. Some say these men are chasing their own shadows, their own dreams. But others—like me—believe them.

I was familiar with the kind of dome structure he described from my travels in Namaqualand. I'd read that in the area around Filabusi, just off the main road from Bulawayo south to Fort Victoria, there were hills described variously as granite domes and polyps. And indeed, in these hills gold had been discovered.

"It's quite possible," I told the others, "it all happened as he said."

After some rest and optimistically "overhauling" the Land Rover, which essentially meant tightening the baling wire around the fender, Patrick, Marcus, and I decided to follow the old hunter's route. However, the civil war situation in this area of Rhodesia was deteriorating by the day. It was really scary.

Some good news was that the Filabusi area—the nearby Doro Mountains—created a sort of natural division of forces in white-controlled Rhodesia. To the south, where we were going, was primarily a white area. To the north, ZANU guerrillas were active, though they had not yet moved appreciably south. ZAPU guerrillas from the west had not yet infiltrated the Filabusi area to any great extent. Thus, the artificial boundaries allowed us some temporary youthful optimism.

The main roads going south to South Africa were sometimes watched by the guerrillas. The white farmers who traveled the roads armed themselves with all types of weapons. One vehicle we saw had a wheel-shaped arrangement of shotguns attached to its roof, pointing outward like a porcupine's quills. If alarmed, the farmer could pull a chain, and the guns would discharge simultaneously.

Eventually, we decided to leave the major roadway whenever we could, entering dry streambeds that were recessed due to erosion. Our dirty white Land Rover was essentially hidden, popping up only occasionally. The beds were cut by an array of old hunting tracks and game trails that we also followed from time to time. In and out of shadows, we moved south, following our compass and maps we had as members of a research unit.

Water sometimes accumulated beneath flat rocks in the sandy riverbeds, so when we drove over them, the front wheels kept going but the rear wheels sank with alarming determination. One morning, with a jolt, we felt the carriage hit sand as the back wheels sank.

"We're stuck?" Patrick asked, waking from a doze. Staying awake in a slow-moving Land Rover when the temperature was 89 degrees Fahrenheit was difficult.

"Very." Marcus's head rested on the steering wheel.

I staggered out from the backseat and focused on the wheels. "She's really in; I mean resting flat on the carriage and glued to the river. We're going to have to work our asses off. This will take hours."

"I'll get the shovels and board from the roof rack," Patrick grumbled.

Two hours later, we were still at it. The Rover kept sinking. At least there were three of us. As we jacked the vehicle up, we pushed in rocks; but they disappeared in a pocket of water that was like quicksand. They were too small.

Humped vultures had noted our dilemma. They stared at us from the bare branches of a nearby acacia tree.

"What's wrong with them, George?" Patrick squinted in the sun, working on the opposite wheel. "You'd think we were lying belly up and raving."

"We are on their sandy plate," I said. "Perhaps you should dig at your wheel with a bit more enthusiasm, so they'll reconsider."

"I'm digging like a dog now. If I go any faster in this heat, they will have me all the quicker. God knows Marcus would leave me here to placate them. Do you know the first thing those savages do is tear open your stomach, insert their heads, and then peck away with their disgusting beaks at the soft parts? You don't have to be a cadaver, either. They'll do it while you're alive, as long as you can't fight them."

"Is your wheel free yet, Patrick?" I asked, not wanting to think about this image.

"The fucking axle is still kissing the sand, as we will be."

I wondered if there were guerrillas in the area, and if they'd see the vultures circling and wonder what was dying. More vultures were landing. The goddamned tree was filling up like bleachers.

Marcus appeared thirty yards away or so, bent over and rolling a huge oblong rock. Patrick and I went to help him.

"Marcus found his occupation," Patrick mumbled.

About halfway there, my Irish friend stopped and whispered between unmoving lips, "Don't move."

Alarmed, I froze and saw a snake some yards in front of us. A long, shiny, slippery green piece of thick rope with a flared hood swayed, coiled and seemingly ready to strike. "Let's run together *now*?"

"No," Patrick—the animal expert—said. "It's an Angolan cobra. He may chase. Better to stay still until he wanders off."

I watched the snake shift from side to side, noting he was in an "up" position. "He's not wandering, Patrick."

"Just don't move. He's more than two yards long. He's concentrating."

"Why are we whispering?" I whispered. "Snakes can't hear."

"Are you sure?"

The reptile had stopped swaying, its beady eyes locked on my lips.

"Look, he's lowering himself. Maybe he will come for us now or go away." Patrick paused. "Just a minute more and we'll find out."

"Find out? Find out what, that he can kill us?" I replied slowly between my teeth.

"Not me. I'm wearing knee-high snake boots under my jeans. He can't penetrate them or strike any higher."

"You bastard . . ."

Marcus looked up from his wobbling rock, swearing with exertion, and saw us. "Oh shit. A snake." He focused on it. "Is that an Angolan cobra?"

"For the love of God, what does it matter, Marcus?" I snarled.

"Aren't they rare?" he asked, continuing to shove the rock forward three feet at a time.

"Shit, Marcus, don't piss him off," I said.

The snake turned toward the advancing rock. It hissed loudly. I leaped away. After a moment of hesitation it slithered off, apparently disgusted.

I took a ragged breath. "Jesus. I should have run."

"No, you would have run toward the Land Rover with him following. He would have likely crawled under it to hide. Then what would we do? It was better we all stayed where we were until he moved off on his own."

"Awwww, shit, you just enjoyed seeing me stand there, Patrick."

He turned away from me, smiling.

With the new rock we jacked up one side of the Rover, rolled the rock under it, and then jacked up the other side and got our board under the wheel. Finally, three hours after the Rover had sunk, we rolled it out of the pool of sand and water and onto one side of the dry stream.

The vultures made strange noises, apparently booing despondently. We climbed in the Rover as Patrick showed his middle finger to the birds while muttering something in Gaelic and then drove away.

Toward evening a running river intersected our dry stream.

"This must be it," Patrick said looking at our map. "There isn't too much else around here, just thorn bushes and Mopani trees."

"Let's give it a try," I replied.

"The prospector said follow this river until we reach an exfoliation dome he called Backside Buff, which from his anatomical comment should be rather unmistakable," I said. "I suppose it's three stories high or so."

As we followed the winding and mostly dry river through a valley, exfoliation domes of various sizes came into view.

"The only thing that matches up to an ass around here is us," Patrick remarked.

Suddenly, as we rounded an elbow in the river, the anatomical feature we sought appeared. A fair-sized crack possibly due to a fault had, over centuries, widened and deepened to such an extent that it vertically bisected the lower portion of an exposed dome. The weathering process had touched up the image by smoothing and rounding off the buttocks. In a stroke of continued creativity, the crack had been progressively widened as it reached the base of the dome. It definitely looked like someone's ass. If that wasn't enough to trigger the analogy bell, there were two round boulders very near the crack.

"Damn," I said, "would you look at that!"

"The boulders look like turds," Marcus marveled.

"Only you would think like that," Patrick said.

"But as a white hunter, he was used to following turds!" Marcus said, correcting Patrick with a smirk.

"Actually, that does make sense," I added, and looked at Patrick.

"Bejesus, you're both daft."

We drove as close to the dome as we could and then walked to the crack. A small but deep stream flowed out of the surrounding ridges, just as the old prospector had said.

Had we really found the place?

"Patrick, you've looked for gold before," I said. "What do you think about this?"

"I read that over the past fifty years there have been many small mines dug into granite hillsides around here, mostly useless. The actual money is made in some nearby bar where the miners go to drown their delusions. However, our hunter found a nugget. That's not an illusion."

Geologically, gold can be associated with milk-white quartz in a fissure or crack within the granite-gneiss bedrock. Over millions of years, heavy spring floods and wind will slowly free bits of gold and quartz and steadily wash them down to the stream. From there, the fast-moving waters will pick up the gold, break it free from the brittle quartz, and deposit a train of flakes and nuggets all down its course.

Patrick looked around, scratching the back of his neck. "If we find some flakes in this area here, where the old man claims he found his nugget, we'll have to backtrack the trail of gold up to the point where the flakes entered the stream. Then we follow the trail into the hills to where it was eroded out of the mother lode."

"So to follow the trail upstream, we'll have to pan for the gold at intervals," I said. "The mother lode, if there is one, will be difficult to locate in those wrinkled-looking hills."

"But we have a nugget for inspiration," Patrick reminded us.

Marcus was worried. "The longer we're here, the less safe we are. Can we speed up the panning process?"

I took off my hat and looked at my friends. "It's slow work, but there are three of us."

We collected our equipment, then hid the Land Rover behind rock hills.

"Where shall we look first?" Marcus asked.

The undulating water would cause the heavier minerals to sift down through whatever sediments were there until they reached the underlying bedrock. Even at that level they might continue to slide along on the smooth rock until some irregular feature stopped them.

"So we look for an elbow in the stream's channel or the downstream side of large boulders," I surmised.

With the molten sun beating down, a shovel over each shoulder, and the heavy gold pans we purchased before we left in tow, we inched across the cool, waist-deep stream until we reached quiet places. Dazzling light played on the water's surface. Patrick bent down and shoveled sand into our large frying pan–size container until it was about three-quarters full and covered with water. Keeping the sand just below the water's surface, he stirred the mixture thoroughly with his fingers and then shook the pan with quick clockwise and counterclockwise motions. Gradually the residue of blond sand and larger pebbles found their way over the rim, and the pan got lighter. It was tedious, hard work.

Marcus leaned over to look. Patrick tapped with his finger to spread out what was left. Several specks of bright yellow flashed in the sun.

"Is it gold or pyrite?" Marcus asked as Patrick fumbled in his pocket for his hand lens, which all field geologists normally carry.

He cleaned the lens on his shirt and gave it to me. I touched the magnified specks gently with my knife point and watched them give way and bend around the steel. They were malleable.

"Yes!" I yelled. "Yes! There's nothing else they could be! Yellow fool's gold—pyrite—is too hard to bend."

"Bless them, bless them"—Marcus danced with outstretched hands—"they are as bright as butter."

"It doesn't mean there's a mother lode anywhere, but it does mean there might be," I cautioned.

Without discussion, we worked our way upstream the rest of the day and found a few specks here and there, just enough to say we hadn't passed the point it began entering the stream. Unlike pyrite, gold shines in the fading light, so we were able to continue panning until shadows stretched far across the valley.

We panned for two days and paid the price for our greed; our hands blistered and sand grains worked their way into the raw areas, cutting even deeper. Too tired and sore to cook in the evening, we ate our tinned food cold under a honey-colored hunter's moon. Africa's baking heat seemed to grow fiercer each day, radiating back from the water and sandy ground. By noon, our energy was halved.

"My whole body feels as if it's decaying, falling apart," Marcus moaned.

Lethargic but still filled with hope, no one replied.

Several hours later that day, as we plodded upstream, the extent of rough and sharp edges on the tiny flakes we found had increased.

That meant they had tumbled less and therefore were closer to their source than the round-edged ones.

"Let's give it one more day," Patrick suggested.

As we worked, oblivious to our surroundings, a twig snapped on the opposite bank. We briefly looked at each other, and I felt the color drain from my face. We slid behind boulders and hid, waiting, lying on our stomachs. Minutes passed under the pitiless sun.

It was quiet. Fear hung in the air. All sound stopped. Any second we could be sprayed with bullets. Something edged out toward the water. I held my breath. It was a zebra.

I heard my friends exhale with relief. The animal bolted when Marcus cursed.

"He's probably very old, couldn't keep up with his herd, and needed water," Patrick rasped.

My heart was still pounding. "It could have been terrorists."

"Yeah," Marcus said.

"Gold mining is a disease, you know, like drinking or gambling," Patrick said. "I saw that in Botswana. It takes control of you, blanks out common sense, and before you know it, you're old, broken, and penniless, with little more than failed memories."

"Or dead at a young age," Marcus added, "as we could be."

Our food was about gone. That evening we ate cereal with dried milk mixed with stream water.

"It's stuck to the roof of my mouth," Patrick said, using his finger to dislodge it. "Do you remember those steaks we had?"

Nobody answered. Smoking our pipes in brooding silence, we knew it was time to move on.

"Perhaps we can come back someday," Patrick said.

"Maybe there is no mother lode; it's just an old hunter's dream."

"I think it might be like trying to return to our youth. In the years to come we might even doubt we really had adventures like this."

"I think we would have to be old and batty to forget this past week," I replied. "Hell, it might be the only thing we remember."

We packed up and climbed into the Land Rover, slowly finding our way out of the bush toward the east until we intersected the main road, which led back to Bulawayo and its airport. The tarred road ushered in the end of the trip for me, and we became quiet. Soon I would be leaving my friends.

I looked at Patrick with his black beard and Marcus with his red one. *Good companions,* I thought. *They are going back to what I've left behind.*

I had left the PRU and, with my student visa expired, could not get back into South Africa. Yet I wondered if I had the staying power to do all I planned, to go back to America and then return to South Africa and continue my studies. It seemed like a lot . . .

We shook hands in the airport parking lot and said goodbye. Our parting was quick. There was no need to linger. I envied them driving away and wanted to be with them. I wanted to return to UCT and continue with my studies, as if my problem with the director's research findings had never happened.

The Land Rover pulled away, heading for a border town and reentry into safe South Africa. I raised my hand in farewell and stared. I wanted them to stop. I wanted it all to go away. It was as though I had to start over again. Did I have the will to complete my dream? Was I being courageous or stupid? Maybe my life would end up like the old hunter's gold story.

Shit. After five weeks of being with my companions, I'm alone again, I thought, still staring at the now-gone Land Rover. Looking down at the tarmac as I walked toward the airport, the sun cast my silhouette before me. *I'm even chasing my own shadow.*

Chapter 19

Equatorial Africa: Nairobi and Mount Kenya

I landed in Jomo Kenyatta International Airport in Nairobi. "Kenya, mysterious Kenya," I murmured as I deplaned onto the tarmac and into the fierce equatorial sunlight like a butterfly coming out of a cocoon. Shading my eyes, I reached the terminal.

It was full of activity. Women walked with pride in long, colorful robes; some had their heads and necks covered in vivid blue, red, and white muslin veils. Men wore flowing *kikois*, the traditional wraparound, topped by stoic, bearded faces and gleaming almond-white teeth. Interspersed with the shuffle of people were cosmopolitan city dwellers in not-so-new Western clothes. Uniformed guards milled about, not much caring if they were in the way. White businessmen, some in light-colored safari suits, frequently popped out of the sea of black. I had read that Kenya was independent and fairly stable. I saw no animosity. I wanted to walk the streets of Nairobi, the famous capital, and see Mount Kenya—more dreams of my youth.

In comparison to most others, I looked scruffy. My hair was now long and ragged and bleached from the sun. My skin was tanned dark brown, and I had on worn jeans, a blue denim T-shirt, hiking boots, and a pack on my back. Regardless, the customs officer, wearing a weary-looking uniform, was all white teeth and smiles. "*Jambo*," he greeted, then asked only the most indispensable questions. He suggested I read the plastic-coated page of immigration rules, which he was surreptitiously cleaning his thick fingernails with.

I climbed into what turned out to be a high-speed minibus—express speed was the norm, it turned out—that rocked and slid as it turned corners. Swaying the whole trip, the bus pulled up with a jolt at a youth hostel outside the city. As fast as I got out, the minibus, churning dust to hysterical levels, sped off. It was like a get-in, hang-on, get-off circus ride.

The hostel was superficially clean. A weighty spice smell, like precipitation, lingered over everything. A few young people lay on the steel bunks lining a larger room. Some scattered travel literature littered the timber furniture. This would be home for a while.

I lay on a bed too, thumbing through the literature but really thinking about my situation. Two hours passed. Then I went to explore the city.

Nairobi was founded in 1899 by colonial authorities in British East Africa as a railway stop on the Uganda Railway. It later burned down after an outbreak of the plague. Nairobi was named after a water hole known in Maasai as *Ewaso Nyirobi*, which means "cool waters."

The famed New Stanley Hotel, a holdover symbol of grand British living, was a watering hole itself; famous people on safari from all over the world, like Ernest Hemingway, had stayed there. They left messages tacked to the *Acacia xanthophloea* adjacent to the legendary Thorn Tree Cafe. White hunters and others sat beneath the tree while smartly dressed black men served food and drinks. It was the old Africa I'd read about, the Africa of hunters, gin and tonic, and evening bush dinners in dark silence gripped by captivating fires. The image swept into me like the air I breathed. It was magic.

Back in the 1950s, the local Mau Mau, mostly members of the Kikuyu ethnic group, began killing white settlers in gruesome ways in the name of revolution against the British colonialist rule of Kenya. The uprising lasted until 1960, when the British army, white settlers, and the local Kenya regiment defeated the internally divided Mau Mau contingency. When I was a teenager, the uprising was so shocking that stories of the viciousness reached my hometown news halfway around the world.

More than eight hundred thousand people lived in stable, modern Nairobi in the late 1970s. People filled the sidewalks, mostly in groups, waving their arms as they talked, picking their broad noses, going in all directions, excited, dressed in both tribal and Western attire. Young girls wore short dresses and tight tops showing long, strong legs and firm breasts. A few stared at me and smiled. Always their bright white teeth, like fresh snow, captured my attention first.

Items were sold on the street everywhere. The smell and taste of fresh Kenyan coffee beans pounded with a blunt stick in a worn wooden bowl was captivating. I drank it and ate spicy Indian samosas, followed by small, sweet yellow corn on the cob. Macadamia nuts, curried eggs, biltong, chicken legs, and meatballs were all for sale. Most people were nibbling on something.

As I walked, I stared at the nearby undulating purple peaks of the hills to the southwest. My God, they were *the* Ngong Hills . . .

"I had a farm in Africa, at the foot of the Ngong Hills," is the first sentence of *Out of Africa* by Karen Blixen. It enchanted me. Her life was particularly out of the ordinary, if not decidedly odd, as she was a Danish baroness. She didn't have to come to Africa; her family had wealth and status. It was 1913, and she discovered a certain freedom in Kenya and saw a chance to accomplish something, so she stayed. Her memoirs said Africa was a place where one accepts the inexplicable.

Blixen and the adventurous pilot Beryl Markham knew each other. Markham was the first person to fly solo from London to North America. At one point during her older years, she lived in poverty in Nairobi. Later, she too was recognized as a great writer. It seemed to me that she and Blixen lived life in flashing moments.

The next day, as I walked the streets, I saw the crowd abruptly veer to one side of the sidewalk. A step later I halted and looked down at a man sitting on the concrete with rags wrapped around his hands. Some of his fingers were gone, and his face was disfigured. He was rotting.

I had never seen leprosy before. His half-hand was out begging. I dropped a coin in it, afraid to touch him. I was shocked. Weren't people with leprosy isolated? It used to be their presence had to be announced by a bell, I recalled reading. Hell, I could have fallen over the poor man. I shuddered. I kept walking, and just down the sidewalk I saw a man sitting with a huge leg stretched out; he had elephantiasis. Another older man had a gangrenous leg, freely exposed for viewing to elicit contributions to his tin can. It was incredible from a Western perspective, but I learned that some *preferred* to have disease, as it gave them an income otherwise not possible; they didn't even seek treatment. No one interfered with the beggars; it was understood they needed money to live. Disease was part of life and had been for generations, and I marveled at the African people's acceptance of it.

The next day, I asked at the hostel how to get to Mount Kenya, some ninety-five miles away on the equator. At more than seventeen thousand feet high, it's the second highest mountain in all Africa after Mount Kilimanjaro, located just to the south.

"You can see it in the early morning before eight o'clock and in the evenings now," I was told. "Most of the time it's in a purple haze, except for a white cap of ice and snow above its timberline. Otherwise, it hides in the unrelenting cumulus clouds drifting by all day. You would almost think it wasn't there."

There was a lodge near the mountain, which offered a camping area with cheap lean-tos. Fellow hikers there advised me that for safety reasons I should not climb the mountain by myself. Two other travelers happened by: a Western man and woman about my age. We agreed to climb together and take the Naro Moru Route to Point Lenana, 16,355 feet high. It was one of several main peaks and named after a chief medicine man of the Maasai people who lived around the mountain.

The medicine man must have been greatly revered, as the local people often thought of the mountain as home to their spirits. Some, like the Kikuyu, with wire coils on their arms wrinkling in the equatorial heat, built their homes with the doors facing it.

In the late 1970s, not all routes to the peaks of the mountain had been defined and climbed. The mountain was a huge old *stratovolcano*—composed of alternative layers of lava and ash—worn down over the last two to three million years. The craggy "nipple" was in fact two volcanic plugs made of more resistant rock than those that surrounded it.

We walked five miles gently uphill before reaching the point where the actual climb began. The mountain can be ascended comfortably in three days, but we decided to try for two. Our goal was simply to climb rather than sightsee. But were we fit enough? We didn't consider altitude sickness, which can occur in a severe form above twelve thousand feet; nor did we consider taking a slow climb to alleviate its effects. We each had only a canteen or so of water.

The lower slopes were very fertile due to volcanic ash deposition rich in iron and other minerals. As a result, coffee, tea, bananas, and other notable cash crops grew all around the mountain, as did massive forests with thick, tall trees. The climate, of course, would change with the elevation. We walked up what looked like a logging road through a hardwood forest at the foot of the mountain. Circling the mountain was a zone in which juniper matured. Growing on a wetter section of one slope were groves of tall bamboo. At the timberline, the trees—and we—were in the clouds. We spent the first night in a climber's lean-to eating meaty sandwiches and other food I'd agreed to buy as my share.

Mostly, our climb was simple and steady, but the trail was faint and seemingly twisted all over the side of the mountain. We just kept going. My heart has the capacity to recover very quickly from exertion, but during one very steep and gravelly portion of the climb called the Vertical Bog, it pounded like a hammer on a rain barrel. Maybe the thinner and drier air exacerbated this; it was something new to me. We all stopped repeatedly, gasping. Crawling on all fours like the residential rodent, the hyrax, was the only way to climb this section. Disturbed rocks, demonstrating our peril, rolled downward a hundred feet, stopped only by the trees below.

We reached the top at dusk and found a small bunkhouse. Before leaving the lodge, my companions had agreed to supply food for that evening. As it turned out, the fellow had selected packages of dried soup. I understood then why he was so thin. It was like drinking a cup of water with a few astonishingly thin, lonely noodles. I could strain it through my teeth. I knew energy was vital. Hell, we were sixteen thousand feet up a mountain!

With nothing much more to do and the night getting colder, we crawled into our bags and slept soundly. Morning came. All was covered in a thick blanket of snow, and more was falling. Everything was white. No tracks, no path, nothing. We had arrived in semidarkness and did not know the way down. I shared the three candy bars I had left.

Shit, I mumbled, *this is not good.*

That was an understatement. Without a trail we could very well end up lost on the other side of the mountain. I had read that some ten climbers a year die on nearby Kilimanjaro and some likewise each year on Mount Kenya. And nobody knew we were here.

We decided to wait a few hours in our sleeping bags, hopeful the snowfall would stop and at least allow us to see better. It did not.

"Well, okay," I said, standing in my shorts and attempting to look out a snow-covered windowpane. "We have to find the path, as it looks as though it will snow all day—and then it will be even harder to get out."

Slowly we climbed down, spacing ourselves out like breadcrumbs, one able to see the other, the last knowing the direction leading back to the hut. After an hour of searching, I found two oddly shaped boulders lying together—boulders I'd seen on the way up. Our path lay next to them. Confident we were on the trail, we spaced ourselves out from that point and stumbled through the snow, looking for another sign. Finally, the full path emerged as the snow disappeared—a relief, to say the least.

Miles later, at the base of the mountain, with nothing around us except nature and a dirt logging road to follow, we took our leave of one another without feeling. We were dizzy from hunger and exertion. I went into the woods by the logging road, wanting to get away. I had been bound up for over a week, and the last two days of tension had only made it worse.

Glad to be off the mountain and by myself, I was a bit giddy but relieved. With that, my body functioned again as I squatted. I stood and admired the truly superb specimen I'd deposited on the woodland carpet—like a giant eggplant. I visualized the forest creatures stopping to stare. They would wonder what magnificent beast had staked its territory there.

Chapter 20

Self-Loathing in Lamu

The dark machine was hissing steam in the evening shadows as if it were in rivalry with its colleagues lined up on adjacent tracks. I was about to head to the old Kenyan port of Mombasa on the Indian Ocean on an inexpensive overnight train leaving from Nairobi. Plaintive siren whistles pierced the station, people yelled in strange languages, and colored lights blinked messages. Anticipation filled the atmosphere.

Pack on my back, I twisted my way down the train's narrow aisle to my designated compartment. The countryside passed by—*clickety-clack, clickety-clack*—as speed was maintained. It was mysterious, dark, and wonderful outside—patches of emptiness broken by great trees, the silhouettes of round huts, shacks, and hills with blinking fires in the distance. Occasional eyes and shapes of forest creatures and men all blended into a land the train punctured and surprised with its great surging headlight. Everything glowed with adventure.

Mombasa . . . It had been the major Portuguese trading center along the entire east coast of Africa in the 1500s. They constructed the prominent Fort Jesus, which served as a trading post and a slave prison, as the area was part of the Indian Ocean slave trade.

From the city's train station, I walked along the flat coastal terrain to the old-town section. The alleys were a window into life, with the smells of spices, standing water, children playing, and discarded things. Friendly Swahilis traded and sold clothing and food. Some presumably Muslim Arab women wore the *buibui*, the nun-like garment of crow black with heads and faces covered. The men wore vividly colored clothing, some with a white turban of cotton muslin knotted loosely around their heads. People brushed against each other as they passed.

Fort Jesus was a main attraction. The fort's naval cannons were capable of firing missiles from three to twenty pounds in most directions. What kind of men had stood on those thick walls, ready to die? What were they protecting with their lives? The slave and ivory trades, it seemed. Somewhere around four million people were sold in the region. The East African trade—the Indian Ocean trade—concentrated on women and children, until the need came for men to work on plantations in the nineteenth century.

In the evening, I walked along the harbor, enjoying the smell of the Indian Ocean. With its dress of seaweed, it rocked in and out, stroking the sand. Seagulls floated in the temperate breezes over the warmest ocean in the world. From time to time the birds swooped down to kiss the water and cry in thanks for the fish.

In the morning, I boarded an old bus for Lamu. Before reaching the island, the bus rolled onto a wooden raft that bobbed from its weight. To my surprise, the passengers, including me, got out to haul on a rope connected to the other side of the large Tana River. We were the engine.

Lamu was small and indeed populated by Islamic Swahili. It had quite a few mosques and veiled women, but some wore whatever they liked. The town's streets were narrow and its buildings strongly made of coral. Patient-looking donkeys were the main transport, going from one palm-and-limestone hut to the next.

The island is very close to the equator. That meant monsoon winds blew northeast from November to February and southwest from April to September. They helped support back-and-forth trade of the triangular-sailed *dhows* that came in from the horizon.

About two miles away from Lamu was a village on the beach called Shela. It had a nice small hotel and, to my surprise, expensive yachts and dhows anchored offshore. The sand was pure white quartz. I passed it one day as I went for a glorious long, hot run on its isolated beach and beyond. I wore only shorts.

After a mile or so I saw someone walking on the beach in front of me. As I got closer, I realized it was a completely naked young woman, short and muscular.

"I have to pass her," I said to myself, looking at her tight behind.

I ran up, clearing my throat, not wanting to startle her, and stopped. "Hi, sorry to disturb you, but I didn't want to pass you without saying something." I tried to look straight ahead and not stare at her nakedness, her muscular shoulders, and large breasts.

"I Mary from Milano," she offered in broken English, and extended her hand.

"I'm George from Buffalo," I replied, stopping fully to take her hand. I tried to look somewhere else. "Nice day for a walk."

"Yes, no rain. No get cold here much. Italy like that but better here."

"You must like the sun?" I offered as we strolled along. "I mean, you're naked." The moment I said that, I knew how dumb it sounded.

"Yes, sun good, and I naked. You like be naked?"

"When I'm taking a shower," I replied, still not thinking. "Well, I suppose it feels good in the sun."

"Why you not naked?"

Not wanting to be rude and not able to think of any reason why I shouldn't be, I took off my shorts, exposing my white skin.

She leaned forward a bit and looked at me. "*Skoose-se*. You big."

"Aaaaaaa . . . thank you."

"You like me?"

"Sure," I said. *What else could I say?*

"You want me?" she asked, her eyes shining.

The idea took hold, so to speak, and I began to get an erection. "Well, I just came out for a quiet run. I think I will continue running before I stiffen up too much." I waved as I hurried forward.

"Stiffen is good," she yelled.

What the hell? Weren't Italian girls Catholic and kind of reserved on this subject? She wasn't carrying any clothes . . . nothing. She didn't have sandals on. *Hell, I'm running and still getting hard.*

Farther down the beach I ran into the cool water and splashed myself. Suddenly, I felt sharp pains up my leg. *Shit! Shit!* I stumbled back on the beach, clutching my leg. I fell to the sand and looked at the welts on my leg. A jellyfish had wrapped a blue tentacle around me up to my knee.

I knew the treatment was applying vinegar and picking the tentacle off with tweezers. I would also need to shave the area to remove pieces that might remain. This knowledge was of no help now on this isolated beach. I hobbled to a beached log and sat smashing the jellyfish with cups of sand in my hand while frantically prying off pieces with a sliver of wood.

My leg went numb. Was this like snake venom? I couldn't walk. Would the numbness abate? Would the toxin take minutes or an hour to spread? I had *no* idea. I wondered if that fucking thing had killed me.

Along came nude Milano Mary, as if on a Sunday stroll. "What problem?" she asked.

"Jellyfish." I pointed to my leg while focusing on her breasts. They pointed at me like cannons. *They may be the last thing I see.* "I can't walk."

She sat next to me on the log. "I massage leg. I know massage. I know Mr. Jelly also from Italy."

"Thank you."

Mary spent several minutes rubbing my thigh and picking away at the remains of the jellyfish with the sliver of wood. She was meticulous and devoted; kneeling on the sand in front of me, she made cooing noises. She was getting turned on. *Part of me is numb, another is getting stiff again.*

"Massage good?" she queried huskily, looking up at me as I began to ease off the log.

"Oh, very . . ."

We had sex.

With Mary acting as my crutch, we later began a slow walk back to Lamu. Needles shot up my leg as feeling returned. We stopped when she felt additional massage was required, or when my hand slipped off her shoulder and onto her hard breasts. I lost count of the number of times Mary and I stopped to rest on the white sands. Someone following our tracks and counting the distinctive sand angels we left in the pristine sand would know. The blazing sun had burned my ass red.

Dear God, what have I become? I had no principles anymore. Still, how often do you get the chance for something like this on an isolated white sand island some fifty miles south of Somalia?

What the hell . . .

Chapter 21

Trading with Xhosas

The time had come to return to Nairobi, catch a plane to the United States, and visit my family.

Seeing my family again was wonderful, but I soon realized my experiences in Africa were alien in my parents' small, quiet house on a peaceful blue-collar street. I owed my family a great deal for the freedom and financial support they gave me when I needed it. They didn't have a lot of money, either. What would I have done without them? The quiet people in the background make the dreams of people like me possible. I loved them but was never good at showing it.

A successful dissertation would make a big difference to my life. It would allow me to substantiate my long studies and justify future adventures associated with my work as a geologist. It was something my parents would understand, my happiness. And it would make them proud their son had risen to such a high educational level. If I failed I'd likely have to return and somehow try to fit in knowing I'd failed; and my employment options would be comparatively limited. I *had* to get back to my dissertation—I knew in my heart I had to keep trying. If I didn't, Joost would win. As much as receiving my degree was my dream, it had now become personal too. Hatred is strong ammunition.

My parents knew I was worrying; I couldn't hide it from them. But how do you explain such a situation? Who would believe me? It felt like such a mess. It was the worst kind of pressure—the unbelievable and unrealistic kind.

And then, a few weeks later, I received a telephone call from Martin. I had been accepted at the University of Natal at Pietermaritzburg (UN-P). He explained the head of the university's geology department—Professor Don Hunter—had sent a letter to Professor NASA at UCT, asking about my problem. It was a form of due diligence. Professor NASA had replied, "In geology, the truth is somewhere in the middle." Apparently, I was no longer 100 percent wrong.

Professor NASA had begun to protect himself. He was no longer in control of me.

The conclusion at UN-P was that there was indeed something strange taking place. It was all still rather unbelievable, but the nightmare *had* happened. Now I finally had the support of key people. A huge weight rose off my shoulders. I was coming out of the shadows.

My parents told me I was smiling more often after that telephone call.

* * *

With a new visa I flew to Cape Town to collect my old car. Patrick had safely driven it back from Jo'berg after our trip. I packed it with personal items, rock specimens, and notebooks filled with data. On the front seat I laid my well-used map. I would have to drive slowly.

Patrick, Catina, and I enjoyed a final Sunday chicken dinner. I showed them my new route, again using a chicken bone as a pointer. Although we laughed, we knew this would likely be our last time together. I promised myself that if I was able to finally publish my dissertation work, I would acknowledge my gratitude to them.

I had the choice of several routes to travel from Cape Town to

Durban, but following the Indian Ocean shoreline along the N-2 looked the most fascinating. The first 150 miles, from Mossel Bay in the Western Cape to the Storms River, is called the Garden Route. That sounded a hell of a lot tamer than the Skeleton Coast or Forbidden Coast.

The locals tell the story of a man named Joseph Wilhelm von Mollendorf, who sailed out of Cape Town, heading toward the Indian Ocean. He was looking for a place to settle. Several people said his ship could have been the Dutch East India's *Maria*, which was lost about that time, in 1788. A wealthy man, von Mollendorf carried his money in the form of jewels and gold coins in a strongbox. The crew put into Plettenberg Bay, where the ship lost anchorage due to a gale and began drifting down the coast to be eventually sunk on rocks. Before that happened, von Mollendorf had enough time to make a raft for himself and his strongbox, and, casting off, he drifted toward the entrance of Ballot's Bay. He rocketed up and down on the berserk waves that smashed their way to the rocky shore near the bay. The raft broke apart, his strongbox sank, and he fractured his arm, but he made it to the shore. On calm days the unsalvageable heavy treasure box could be seen wedged in the rocks below a hysterical, deadly current. The lost box remains a kind of folktale that some swear is true.

About six hundred miles east of Cape Town on the N-2 sits peaceful King William's Town. Nearby is the port town East London. It marks the beginning of the 150-mile-long Wild Coast, named for the large number of shipwrecks there. Just north of King William's Town, the N-2 crosses the Great Kei River. This waterway forms the southern border of what was the first independent, apartheid-driven homeland for some five million Xhosa natives, named the *Transkei*, which means "the area beyond" (i.e., beyond the Kei River). Prior to this, over the past two centuries, the Xhosa had lived in the broader southeast portion of South Africa. The Transkei was isolated and peaceful.

Travelers are met with a variety of greetings like, "How are you?" The traditional Xhosa greeting is, "*Inkaba yakho iphi?*" which means, "Where is your navel?" After giving birth, the Xhosa mothers are required to bury the afterbirth and umbilical cord near their village. The answer to that greeting conveys where the respondent comes from, what his status is, what clan he is from, and so on.

I drove north, arriving in the port city of Durban, in the province of Natal, a day or so later. It was named by Vasco da Gama, who anchored in the bay of Natal in 1497 on Christmas Day: "*Natal*" in Portuguese means Christmas. (Following the 1994 elections and the formation of a black-led government, the province would become KwaZulu-Natal in honor of the eleven million native Zulus living there.)

I turned inland at Durban and followed the N-3 west fifty miles to the town of Pietermaritzburg, where the UN-P campus was.

In 1979, the year I arrived, the University of Natal had two campuses—one at Durban, the other in Pietermaritzburg—and was one of the three best universities in South Africa. I was happy to be able to continue my studies, also receiving a grant as an incentive.

Pietermaritzburg was founded in 1838 by Afrikaners who traveled there by ox wagon. Following a fight, the British took over the city in 1843.

The city is regarded as the birthplace of legends. Some years prior to my arrival—seventeen to be exact—and not far away, Nelson Mandela was arrested and began his twenty-seven years in prison. In 1893, Mahatma Gandhi was thrown off a train in Pietermaritzburg, which prompted him to make the momentous decision to stay in South Africa and fight against the nationwide racial discrimination of Indians.

The university's two campuses became a major voice in the fight against apartheid, and UN-P was one of the first universities in South Africa to accept black students. In this quaint place with historic buildings, a multitude of aged trees, uncongested roads, and a famous botanic garden, I immediately felt comfortable.

Martin welcomed me. We discussed in detail the metamorphic petrology and structural geology of the area integral to my dissertation. He knew the subjects well and showed great interest in what I said. Finally, he shook his head in amazement at understanding the problem. My days at UN-P would be busy. I had about eighteen months to finish my dissertation. And I looked forward to it.

I got right to work. It was my custom to work on my dissertation studies until two or three in the morning, as I was a night person. One evening, I sat in my office with my window fully open to a warm summer breeze. It would rain soon; the feeling was in the air. A sizable ant flew onto my desk. I had read that from time to time a large, fat-looking queen ant flies off by herself to establish a new nest. Could this be one? *Amazing,* I thought, *that a rare queen ant would land right here in my office. Finding a new home for herself and her people is an awesome responsibility.*

I gently maneuvered her onto a piece of paper—she wasn't aggressive—and put her on the outside window ledge before, in an act of unnecessary deception and considering her size, I hit her with a rock. She did indeed splatter. The truth was I didn't like ants; too many smaller, more aggressive varieties had bitten me as I slept in the bush.

Another "queen" entered. Two? I stared in disbelief at such odds. In a moment more came in, and then more and more. "Shit!" I said, slamming the window closed. These weren't queens; they were all exactly the same size: huge. Flying ants, they landed all over the place. I began swatting them. Thousands of golden clouds swarmed outside in all directions, obscuring streetlights. It was madness, an attack from space.

I learned the next day that they flew into car windshields and left thick, oily spots. Smacked and run over, they left giant puddles of grease, making the roads slippery. As it turned out, just before a rain, flying ants leave their nests to establish new ones. *Millions* of them look for a dark, wet area that will allow them to survive when the burning sun comes out. It's a frenzy of wings—all beating mightily at their one chance to perpetuate life.

The black people eat them; the ants are full of protein. To catch them they sometimes make a noise that sounds like rain by brushing straw on a pail over the nest. The ambitious ants are fooled into flying out into a waiting basket. I liked that trick.

* * *

To pursue my studies, I drove south several times to the Transkei. Certain rocks there were of igneous origin. Thinking they might help me visualize my metabasite rocks before they were metamorphosed, I packed a Land Rover belonging to the geology department and stopped along the way at a small town called Tabinkulu. I became friends with Henry, the only white trader in the town; he owned a large barn-like store where I bought supplies.

The dirt track leading to Henry's trading post was just two parallel red ruts or, in the rainy season, two red streams. A few old acacia trees rose sporadically out of the surrounding open fields, which were covered in tall elephant grass.

From my Land Rover I could see distant paths broadly zigzagging through the grass-covered, round hills that led to the Xhosa mud and thatch-roof huts (also referred to as "brown nipple" huts)—always on hilltops. Several old and bent-over women were slowly carrying the ubiquitous four-gallon paraffin tins that served as water vessels. They balanced these on their heads up the steep trails, looking a bit like snails. Life hadn't changed in hundreds of years here. *It was Africa.*

After a few miles the track descended into a forested valley. I caught up with an ox-drawn sleigh with woven sapling sides moving lazily with the rhythm of the powerful animal. The driver, a bare-breasted pretty black Xhosa girl, whipped the sullen beast with a stick to the side of the track. She smiled at me with bright white teeth as I raised my hand and passed.

Because it was Saturday morning, she would be taking sacks of mealies, corn, to Henry's store. I wiped red dust from the rearview mirror and gazed back at her. Without Henry's gristmill, she would have had to grind the corn herself on a rock with a smooth river stone.

The track widened nearer to the store, where I parked. Several groups of women sat discussing their purchases. Another was trying to lift a newly purchased coffin made of uneven boards onto her head. Several shoppers and I helped lift and balance it. It was heavy, and I knew it contained whatever additional items she had bought. She would probably carry the coffin a considerable distance, possibly up one of the hill paths to her home where someone lay either dead or dying.

The one-story trading store was twice the size of a large barn. Higher-priced items, like the varnished plywood coffins with shining brass handles, hung between the slowly rotating overhead fans near the ceiling and distant from the edibles. "It helps keep the store a bit cheerier," Henry pointed out. More in demand were the clothing, bolts of bright-colored cloth, enamel kettles and cooking pots, bars of blue washing soap, beads, bottles of cooking oil, tobacco, paraffin, and staples like sugar, milk powder, flour, and salt, as well as rows of tins and boxes of this and that lining the walls and floor.

The store also sold hardware, like various gauges of wire used to construct the beaded bracelets that covered entire arms, with only elbows visible. In Xhosa society, bracelets are a status symbol. Hands and fingers are left undecorated, available for labor. I did, however, notice an older lady who had adorned several of her fingers with the pull-off tabs from soft-drink cans.

For some women, such decorations on the little fingers of their left hands would be impossible. It seemed to be a custom to chop the digit off at the joint when they were born, with the expectation it would prevent ill health in later life.

Henry spotted me through the busy mass of shoppers—some with babies strapped to their backs—and invited me into his office. We were the only whites around for miles and about the same age; both enjoyed hearing about each other's life. As a youth, Henry's father sent him to an all-white school elsewhere in South Africa, and since returning to run the store, he didn't see many people of his color. He offered me a whiskey and apologized that there was no ice.

"A snake crawled into the generator room last night when it got cool outside," he explained, holding a glass up to the light and frowning. "It was attracted by the warmth of the motor. The servants are afraid of it. They came out screaming this morning, and I haven't had time to chase it." After pouring a whiskey for himself and for me, he sat down and crossed one leg over the other.

"What kind of snake?" I asked.

"Spitting cobra. If the venom hits you in the eye, which is where they aim with terrible accuracy, you're blinded immediately. The worst part is they sometimes fake being dead. You can beat the hell out of one and think it's gone—even pick the damned thing up, limp and hanging, shit dragging from its ass—and wham! All of a sudden your eyeballs are full of snake spit, and you can't see."

"We can do without the ice, Henry. Cheers," I replied, raising my glass, then feeling the raspy liquor coat my throat.

"Cheers, George."

I sat quietly for a moment, reaffirming my suspicion that the whiskey was cheap when my tongue dragged over the roof of my mouth like sandpaper.

Henry put his feet up on his paper-laden desk and stared out his dusty window. I looked out, too, as the oxcart I had passed stopped under the window. Another girl had joined the first. A light-skinned baby was sucking at her breast as she shooed flies away from its running nose.

"A lot of flies," I commented without thinking, distracted by the women's breasts.

"It's dry now," he said. "The flies need water to drink, and I guess snot is as good as anything."

"Are you watching snails, Henry?" I asked, nodding at a crude aquarium in a corner of the office. Filled with red dirt and vegetation, a number of snails had attached themselves to its unevenly matched glass walls.

He turned around in his chair and focused on the aquarium. "I suppose you could put it that way. About a month ago I saw one creeping up the garden wall and watched it crawl over its own silver slime. It was like laying down your own rug before you. I was fascinated. I decided to put some in the tank to watch them." He leaned forward on his elbows and squinted at the aquarium. "Their eyes are at the end of their horns, and I actually saw one put one of its horns down its gullet. I think it wanted to see what it ate. Why it didn't look before it swallowed, I can't figure out. Maybe what it ate turned out to be bad."

I watched the snails while Henry went on. "What else is fascinating is that I reckon every one of them has had a go at the other. They seem to have sex indiscriminately and even simultaneously with both sexes. Anyway, their approach works. I've no shortage of snails."

He drained his glass and stood. "I'd best take a walk through the store." He handed me a box of assorted items he must have received as trades. "Take a look at these. I'll be right back."

"I'll find some information on snails when I get back to Durban and bring it to you," I called out as he left.

The box contained baubles: intricate bead necklaces and bracelets painstakingly sewn together; pipes; homemade knives; a few short spears; and at the bottom, a distinctive purple rock I recognized immediately as bornite. Henry knew that as a geologist I'd be interested in the rock. I suppose he wondered what it was and if it had value, which is why he slipped it into the box.

"There's an *igqirha* outside who wants beer," Henry said as he returned and sat down with a sigh. "For free."

"A witchdoctor? What did you tell him?"

"That the beer is in the generator room, and he could help himself."

I laughed. "Did you warn him about the snake?"

"I told him and asked if he could remove it. He said he would go home and brew a mixture that would make it leave in a few days."

"It will get hungry and leave anyway."

"Yes, or I'll get bloody sick of warm whiskey and chase it myself. Did you find anything in the box?"

"Some very nice beadwork."

"Anything else?"

I held the bornite in my open palm. "This is an ore of copper."

Henry raised his eyebrows. "Is that so?"

"Miners call it 'peacock ore' because of its color, which means it contains a high percentage of copper. Perhaps it's from a vein that leads to a larger deposit. Where did you get it?"

"From the local old *inkosi*. He offered to trade me that for a large three-legged cooking pot. Is it worth the trade with the chief?"

"There must be more, but the price of copper is always low. You would need a lot, which is possible here due to the geology. Best tell the old chief he has to take you to the spot in return for the pot. If you'd like, I will go with you."

I turned the bornite over in my hands and had no problem imagining it as part of a large deposit of copper.

"Okay, let's have another drink to celebrate hope."

I went out into the store to buy some food and a new shovel, as my old one had broken when I used it to pry a rock loose. As I left, a throng of men and women dressed in wraparound clothing as well as odd pieces of Western attire, including men's hats, had gathered just outside the front door. They were in a commotion about something and raised a light cloud of dust.

At first, I thought they were just staying in motion so the ever-present flies would be less likely to land. Then, peering into the crowd, I saw an old woman at their center, sitting on a wooden box. She was holding her head and staring ahead, glassy-eyed.

Suddenly, she removed her hand from her head, and the crowd gasped and groaned. What looked to be a ten-penny nail head protruded from the top of her partially shaved skull. She said a witchdoctor had driven it in. She also had a tuft of goatskin tied at the back of her neck, a sign that someone, probably another witchdoctor of lesser importance, had tried to cure her of the unpleasant effects of the nail.

Henry heard the moans from the crowd and came outside. I had discreetly retreated from the group, not knowing how they might react in their agitated state. Like a nervous wave, the mob parted for the white trader. Silence prevailed.

Henry looked at the head of the old *umakhulu*, grandmother, for a full minute. No sound . . . the calm before a storm? I calculated my chances of reaching the Land Rover should it all go badly.

Henry turned and faced the Xhosas. I could tell he knew that if this didn't turn out well, the crowd could get very excited and whip themselves up to a point of damaging his store. He recognized the inkosi who wanted the three-legged cooking pot and spoke directly to him, knowing the people would listen to their chief.

"If you think it best, Inkosi, I'll take old umakhulu to the hospital where Xhosa and white doctors will help her. When I return, I will fill your cooking pot with homemade beer, and you and your people can discuss the actions of the igqirha."

The inkosi understood. Part of the deal was that he would get the three-legged cooking pot he wanted. He probably also realized that the tuft of goatskin tied to the old grandmother hadn't cured her.

"Take old umakhulu," the chief commanded.

"*Hiiiiiiiii . . . Hiiiiiiiiii . . .*" the crowd chorused in approval.

Henry gravely nodded acceptance of the inkosi's decision, and the group dispersed.

I watched Henry walk toward his car with the umakhulu.

Wanting to avoid any problems, I went briskly to my Rover and drove ten miles or so to the rocks I'd come here to investigate. A few days later, I returned directly to the university.

Several weeks later and curious about what happened with Henry and the umakhulu, I visited him on my way to my field area. He was staring at the closed cash register and glad to see me. Xhosa customers strolled up and down the aisles, buying nothing but fingering everything as if they had never seen it before. A can of soup would generate a discussion.

Later, as Henry and I chatted, I asked him about the old grandmother.

"The hospital told me the witchdoctor had scraped skin away from her head and rubbed in a powder that caused irritation." He looked up at me. "An enlarged scab had formed from excessive scratching. It looked enough like the head of a nail to scare the doctor pretty well at first. They explained this to the woman, but she didn't seem to understand. She just sat there, staring, apparently convinced that a nail was in her head. Last I heard she hadn't come out of it yet. Perhaps she never will."

I shook my head and watched a chicken enter through the open door. One tentative foot at a time, it advanced on the concrete floor, watching people sidelong. It pecked its way down an aisle, following a trail of spilled mealies.

"Not much the doctors can do when rural people become convinced of something," he continued, now stacking Rhodesian tobacco on a shelf.

"Poor old lady," I said. "And did the chief get his cooking pot?"

"Yes. Then he wanted an additional three blankets to take me to where the bornite came from."

"What did you say?" I asked a bit amazed.

"What the hell could I say? I agreed to give them to him. Then, all pissed off, I went into the generator room with a broom and fenced with the fucking snake until it slithered out the door. I filled a glass with ice and whiskey and sat down to watch the uncomplicated snails."

Hearing a noise, we turned to see the chicken fluttering its wings, apparently as its red-rimmed eyes spied its own reflection in a chrome milk bucket. Henry excused himself and went for the chicken, which, sensing danger, began to run, surprisingly agile, between the customers. Although Henry tried to maintain his dignity, it was difficult. He returned, somewhat flushed, with the chicken struggling under his arm, and continued our conversation as if nothing had happened.

We made plans to take the inkosi to where he or somebody in his tribe had found the bornite. "Let's try, Henry," I said. "It's one of those things you just have to do or you'll always wonder."

Chapter 22

Guano over Everything

On the appointed morning, Henry and I drove in his battered old Land Rover to the chief's hut. The day was already warming up. The chief, still wrapped in a blanket, climbed in the front seat—the only seat—between us. He smelled badly and took up a fair amount of space with his large beer belly.

"Oh shit, Henry," I mumbled, holding my breath, "open the windows."

"Any beer?" the chief asked, reminding Henry of the complexity of the arrangement.

"Sure." Henry pointed to a case in the back among other supplies and suggested we all have one.

For miles we jostled along unused dirt tracks mantled with tall ferns and patterned in gold and shadows until we reached the crest of a forested ridge. "Magnificent scenery," I remarked as I looked down at the undulating hills and valleys. I turned to Henry. "Have you traveled this way before?"

"Years ago with my father, just as he had with his dad."

"How did your grandfather get here?"

"My father said he was shipwrecked, but he always said it with a smile, so who knows. He may have come here to get away from the law or the British army."

"That's a bit mysterious."

"It's not so mysterious here. My father must have thought it best not to discuss it."

"Ever thought of leaving here? Perhaps moving to a city?"

"Yes, but this is my home. What would I do somewhere else? I don't have much formal education. Christ, I'm watching snails." He swerved around an ant mound.

"Nothing wrong with you, Henry. You're what's called street-smart. Here I am, educated with nothing but a pile of books, a dream to get a degree; and you own a big store."

He took a long pull on his beer. "A store I might lose because I'm a white man."

I glanced at Henry and the chief, their profiles bouncing in unison beside me.

"Hell," Henry continued as the chief belched and passed gas, "the blacks and whites in South Africa are only just beginning to understand and work with one another. Not all this business of recent sanctions and forcing confrontation makes sense to me. How can people from other lands understand what we ourselves don't? Give it time, and apartheid will crumble of its own accord."

The chief, silent to this point except when he wanted beer, motioned for us to stop. Alighting, we all lined up at a tree like dogs marking their territory. The chief struggled with his fly.

"What makes you think apartheid will collapse?" I asked.

"People need each other to survive. They give and take until balance takes place. You know, before the whites arrived in Africa, the Africans controlled trade themselves for over ten centuries. At one time they supplied the world with most of its gold." He repositioned himself. "Nature speeds things up, too. Eventually all will breed together. Look at the three million colored population—they are part white and part black or something. If snails can figure out how to mix, why can't we?"

Henry and I walked back to the Rover. A minute later the chief finished climbing in between us, struggling to close his fly. Henry offered words of encouragement that caused the chief to grin.

In a moment we were weaving down the track. Swallowed by the thickening forest, Henry wrested with the steering wheel and cursed each time he ground the gears together. He cut his lip on his beer can when we bounced over an ant mound, and blood dribbled on his shirt. Now he drank out of the side of his mouth.

We bounced together: I, the battered trader, and the stoical chief. *Here we are in Africa,* I thought, *going down some nameless track. Life is interesting.*

The chief stared ahead with heavy-lidded eyes and swayed with us. I drained my warm beer and tossed the can with the others on the bed of the Rover. "Do you think it's much farther, Henry?"

"Christ, I hope not. There's a spring in the seat that's about to enter me."

"What about you, George?" Henry asked as we drove on, holding on to the dashboard from time to time. "You're a long way from New York. Are you looking for money? Trying to find your pot of copper?"

I drained the remains of my beer before replying. "I'm following my dream and want to contribute to science. But, sure I'd like to find a copper mine," I said quietly.

"It sounds partially noble."

I imagined every great copper mine I'd seen. All of them found by fellows like me—guys with a stack of wrinkled maps, a Brunton compass, and a rock hammer. What they had more than anything else, though, was luck. In a few years the large deposits still to be found in the world would be discovered by scientific methods. The old prospecting days were over; the remaining deposits buried too deep in the earth with no recognizable surface expression for the passing geologist.

Henry was about to say something when the chief abruptly leaned in front of him and blocked his view. The Rover hit another rock, and we bounced high. Henry bashed his lip again.

"For Christ's sake!" Henry swore as he tried to regain control of the Rover.

"There!" The chief jabbed a finger toward the foliage.

Tree limbs beat the windshield and doors as Henry frantically pumped the weak brakes.

"Time to walk," Henry said slowly as he lifted his rucksack and tentatively touched his swollen lip. It had begun to bleed again.

A mile later, the tottering chief, now without his blanket, pointed to a sheer wall of rock covered with very heavy vegetation. "Rock from there," he proclaimed.

Henry focused first on the innocuous wall and then on me. "Your move," he said, scratching himself.

I dug around in the foliage and found pieces of copper-bearing minerals scattered here and there. *What is this? I* wondered. *Why are broken pieces lying around?*

The chief interrupted my dreams by asking for another beer. "Jesus, Henry. Does he have any notion of what we're up to here? Does he know what a mine could be worth?"

Henry grinned broadly. "He's thinking in the short term—the three blankets he's now almost earned and the immediate relief of his thirst. It's day by day here. You, white man, think in the long term."

"Pass me a machete, Henry."

"My pleasure," he replied, seeming happy there was only one available.

I hacked for some minutes, peering ahead, and finally reached the base of the rock wall. I found a man-sized cavity in the wall. It was a tunnel: an old prospecting tunnel!

"Get a flashlight!" I called through the vegetation. "I'll bet no one has been inside for a hundred years. It's a forgotten shaft. Henry, stay outside. The chief and I will go in."

Henry motioned to the chief, who began to retreat.

"What's wrong with him?" I asked.

"He doesn't know what's in there. Actually, neither do I."

"Hell, I don't, either! So what?" I was thinking about money. Thinking more, I realized there could be something else in the tunnel. The cavity would make a fine house. "All right, I'll go first. Tell the chief to follow. It was part of the deal to show us."

The small, round-roofed tunnel, with a single-rail track for taking out rock, widened after about forty yards in.

"Shit, this really goes into the mountain," I turned to say to the chief. I saw that he had advanced only half as far as I had. About that time I became aware of shuffling noises above me. I shone the light upward.

I was not prepared for what I saw: hundreds of very restless, ugly bats about the size of starving pigeons, angling upside down as only a bat can. I dropped low automatically. Their radar was impeccable, but mistakes can be made; and I wasn't happy about the idea of being struck in the face by a misguided or aggressive bat. I decided to chase them toward the entrance rather than chance an uncontrolled situation. I protected myself as best as I could—waving the machete in one hand and the flashlight in the other while quickly shuffling a few feet toward the now-distant entrance.

That was all it took. They unhooked themselves and uttering shrill squeals and raining white guano, they took off on the obvious course, away from my flailing arms and rotating light—toward the entrance.

Unfortunately, I had forgotten about the chief, now forlornly frozen in the dark passageway halfway between me and the entrance. In the distance I caught a glimpse of his blinking white eyes in my flashlight beam as he tried to grasp the situation. Too late. The disturbed bats, their cries echoing, closed in on him with frightening speed.

"*Howwww . . . howwww . . .*" he wailed as he ducked and stumbled toward the entrance, his arms overhead and his stomach bouncing.

The bats whisked by him on all sides at astonishingly close margins. Henry heard the commotion and came rushing into the entrance. His frame filled the small opening. The lead bats, bright-eyed with their radar fully honed, received signals that their escape was blocked and came screaming back at the chief.

"*Howwww . . . howwww . . .*" he continued, louder, as they pelted him with guano again. He staggered back toward me.

"Get out of the entrance!" I yelled to Henry, who by now needed no instructions as some of the bats forged past, brushing him.

Once again I raised the flashlight and waved. The bats, now hysterically confused, responded immediately as did the chief. All raced for the entrance.

The tunnel emptied of bats. The chief lay outside, coughing and swearing as Henry tried to calm him.

"White boy crazy! White boy crazy!" he wheezed.

Scanning the walls with my flashlight, I quickly found a six-inch-wide seam of bornite. Hurrying, I followed it farther into the mine. "Take me to the mother lode," I said aloud, my fingers tracing over the seam.

"What did you say?" Henry's voice echoed down the tunnel.

"The copper," I replied mechanically, completely engrossed at the sight of the seam. The light from my flashlight brought the beautiful ore to life as the morning sun might.

I followed the seam with my fingertips as fast as I could for another ten feet, where it decreased to four inches. *Not good. Not good at all,* I thought as I stumbled along. I scanned the wall on both sides of the tunnel. A few feet farther, the seam abruptly wedged out to a pencil-line thickness. I stared at it blankly.

"How's it looking?" Henry called.

"It's wedged out."

"You mean disappeared?"

"No value, Henry," I called back to him as I stared at the wall. "No market value at all." The seam ended in nothing. "*Nothing,*" I repeated out loud. My chance for wealth disappeared into the mountain. *Ahhhh, shit.* Crestfallen, I turned to slowly walk out of the tunnel.

As I neared the entrance, I could see his motionless silhouette and then the disappointment on his face.

Without speaking, we returned to the Land Rover, where the chief, stippled with guano, waited, drinking yet another beer.

"You're starting to look like a white man," Henry said to the chief, who looked at him without comment.

We drove back to Henry's store.

Once there, and refreshed after sleeping the whole way back, the chief asked for his payment of three blankets.

Henry gave them to him and explained that there wasn't enough of the purple rock in the old mine. The chief thought about this and then said that he knew where there was a very big hill of purple rock. If Henry would give him some brass wire, a shirt, and a pair of shoes, he would show us where it was.

Chapter 23

The Procession: Baby's Coffin in Red Clay

A year and a half passed quickly. During that time I'd visited my field area in Namaqualand with Martin, my new advisor. I looked at countless thin sections, read papers on other scientists' parallel work, and plotted hundreds of graphs using the chemical data I'd collected at UCT. I also began drafting chapters of my dissertation. And before I knew it, just six months were left of the four years allocated for study. It was time to finish my presentation.

I began working twelve hours a day and more, seven days a week. My life rolled on without variation, like a clock ticking without numbers. And then, one day, helped along by Martin's unwavering encouragement, I finished my dissertation. To this day, Martin remains one of the most instinctive and intelligent men I've ever known.

It was 1980. The title of my dissertation: "A Geotraverse across Namaqualand, South Africa: The Petrology, Geochemistry, and Structural Relations of a Proterozoic High-Grade Metamorphic Terrain." It was 203 pages long and bound in book form.

All the data collected during my time at the PRU pointed to the same major scientific conclusion. That is: *Namaqualand is the site of an extensive Proterozoic granulite terrain from which the bulk and mineral chemistry of the mafic specimens show no consistent variation in the concentration of* Al_2O_3*,* Fe_2O_3*, FeO, MgO, and CaO along its forty-five mile length. The evidence showed that the coastal five-mile lower-grade zone adjacent to the granulite zone is the result of later retrogressive metamorphism rather than progressive metamorphism, as previously suggested in the literature.*

It said Afrikaner Joost was dead wrong.

The university submitted my dissertation for external review to three eminent, widely published geology professors based in Britain, Australia, and South Africa. Martin had selected them because of their knowledge of metamorphic petrology, geochemistry, and related structural geology as demonstrated by their publications. I knew I would have to wait at least a month before I received their responses. I also knew it would be a long month.

Martin felt confident that my work would be accepted. But after all the negativism I'd been through at UCT, I was not so sure. In spite of Martin's reassurances and calm voice, my nerves were ragged as I awaited their feedback.

* * *

During the waiting time, I was offered a lecturing position at another university. It would start in a few months. To take my mind off waiting, I decided to return to the Transkei, look for minerals, and wander a bit in the rolling countryside.

A few days later, I found myself at the top of a golden grassy hillside, one in a long line of foothills in a very rural area of the Transkei. Surrounding me were elongated and round large white granite boulders nestled within a meadow of the earth's tresses. It was an early, clear morning, a suitable place for looking off into the distance over lone green trees and rolling valleys.

Three teenaged native boys joined me in digging. I was searching for minerals that had formed millions of years before but might lie just below the earth's surface.

South Africa has the lion's share of the earth's mineral reserves: nearly 90 percent of the platinum metals, 80 percent of the manganese, more than 70 percent of the chrome, 45 percent of the vanadium, and 41 percent of the gold. Why is the country so mineral-rich? No one knows, but a running theory is associated with the unusual presence of very ancient rocks older than those I'd studied. Given such rock has survived, this suggests that subsequent erosion and tectonic activities may have been less than those affecting other areas of the world where such ancient rock did not survive. Therefore, vast mineral reserves also survived here.

Such wealth could be under my feet. I looked around me and once again saw earthen *rondavel* homes nestled here and there along the steep grassy slopes characteristic of the Transkei Wild Coast. On average they were twelve feet in diameter and seven feet high. They were like the tops of huge mushrooms and they stretched into the distance. Through shimmering heat waves, the trodden red-clay earth surrounding each hut looked like a target. A characteristic circular fence made of thorn branches and grotesquely shaped tree limbs enclosed each dusty earth patch, creating an open yard-like area.

African women were up before the sun rose, sweeping these courtyards, not wanting their neighbors to think them untidy. Uninvited people were not to enter the yard. That was as bad as barging into a Western home unannounced. Inside the one-room huts, each member of the family had a mat to sleep on. During the day, the rolled-up mats hung suspended from the roof alongside woven grain baskets and similar items. A few huts had Western furniture, such as a metal table or a plastic chair, depending on whether the head of the family worked in faraway cities and mines.

Some hours passed before a gray and bearded village elder, his walking stick in hand, slowly zigzagged his way up our hill and sat down beneath a nearby acacia tree. Although the man was dressed in rags and seemed to lack status, it was important that we show respect to his age. He had come to make sure his share of the young boys' wages was paid. Tradition demanded that. Otherwise, the sun would have confined him to his home in his chair—a crate that he moved around his hut to catch the shade. At midday the sun forced him to go inside and sleep. Life could not be hurried here.

A sparrow-size *titihoya* flew past us. Perhaps its thorn-tree nest was nearby. I listened to its desolate cry, so characteristic of the African veld. One boy instinctively grabbed a fragment of rock from our diggings and hurled it at the feathered creature, killing it. He hastily collected handfuls of dry grass and twigs and laid them in an already sun-heated, saucer-shaped depression in one of the white granite boulders. Then he set fire to the collection with the bird on top.

A minute later the boys laughed at the size of the portion each received, their white, straw-polished teeth dominating their faces, capturing all my attention. The elder had respectfully accepted the largest morsel from the hunter.

The remaining ash and drops of blood marred the brilliance of the white boulder. It would be a monument to the bird until the rain washed it clean.

I looked away to see a group of Xhosas on the next hill. They shuffled in single file around knots of boulders constricting their path as they wound downhill. The rope-thin path led to the valley between them and us. I could see them clearly, a funeral procession. Their meager clothes reflected the color of their land.

My young diggers glanced across the valley and then returned to their labor while the old man stared. The procession carried a child-size crude wooden box, the best they had. Splinters would surely remain in their hands, but they wouldn't notice. The hole they had dug and were now moving toward was similar to ours but closer to the dry valley streambed. When the floodwaters laden with blood-red clay came, the earthen tomb would be secure.

The Xhosas gathered around the grave. A single caring voice rose from the silent others. It penetrated the valley's calm.

The mourners lost little time. They quickly filled the small hole with clay and rock fragments. The rich red earth, scorched by the sun, lashed by the wind, and shriveled by the drought performed yet another service. Except for two figures that stayed behind, the quiet procession trod back up the hillside.

The child would be in his grave while the land above him would be torn and cleansed when the spring rains came. Life would return to the way it was, just as it had after the bird's passing.

Sitting and staring into the distance, my knees up and my arms folded across them, I thought about my life. The early stage was over, sooner than I believed it would be. It gave me enchanting campfires and dreams to follow. *Would there be many more?*

Chapter 24

1979: The Doctorate and the Witchdoctor

After several weeks, Martin informed me that the first response to my dissertation had arrived. He handed it to me, as I tried to interpret the expression on his face.

I riffled through the pages. Several contained comments on spelling and word corrections and suggested figure and graph caption improvements. The annotations were a normal procedure designed to instruct me in ways to make the document better. I sped through them all, my heart pounding as I took in every other word, looking for the punch line.

Finally, near the bottom of the final page was the magic statement: "I have no objections to the degree of doctor of science being awarded."

My mind exploded with happiness, and then I brought myself back to reality. I needed two of the three reviewers to approve my dissertation if I was to get the degree.

The tension continued. I began to go for runs. No matter what I did, regardless of Martin's unstinting support, nothing put my mind at ease. I suppose I was somewhat paranoid thinking that somehow I would be defeated. I decided to do what had worked before—I got into my Land Rover on Sunday morning and took to the road.

My destination was a valley not far from Durban on the road to Pietermaritzburg, known to the Zulus as eMkhambathini, the place where the giraffe thorn—a type of acacia tree—grows. God created the valley, say the Zulus, when he became disgusted with the world and crumpled it in his hands and then released it when he changed his mind. The result was a rugged valley with high, sharp peaks rising to breathtaking heights and a deep, plunging valley formed by the Umgeni River on its travels to the ocean. The valley looks like moss growing on gigantic pieces of broken, sharp glass.

The land, like many such places in Natal, is a mystical place. It gives rise to superstitions and myths. The isolated people living there believe that the mountains, rivers, waterfalls, and rocks have spirits. It is the center of the universe, they've concluded. God had evidenced that belief by creating crumpled mountains. Beyond the mountains, the land is flat, with no significance. The residents of eMkhambathini are convinced they have the best part of the world.

On the day I visited, I passed Zulu villages and colonial dwellings in the valley. Tourists stop at some of the villages. Most remain secluded, however, as access is difficult and strangers are not encouraged.

I drove my Rover slowly along a road obviously not much used, and came on a clearing in the forest. A *sangoma* lived there—a practitioner of herbal medicine, divination, and counseling. I saw him sitting on a crate in the sun, dressed as only as a sangoma dresses. He looked about ready to fall asleep—maybe he already had.

A number of mud huts with straw roofs surrounded him. They were beehive-shaped, one for each of his wives, some of whom were outside watching me as I drove. An assortment of goats was busy sorting through garbage, and several chickens strutted about, twisting their heads.

I decided to stop. Such men are normally intelligent, sociable, trustworthy, and curious, so I was hopeful he wouldn't mind my intrusion.

"Hello," I said, introducing myself. A pungent herb aroma emanated from the hut just behind him.

He looked to be about thirty. He stood, and we chatted a bit above the noise of the goats. He would expect a little money after our talk.

He wore a small tartan blanket around his shoulders that had been washed and exposed to the sun to the point of blandness, and a loose-fitting squat leather dress. Beads were everywhere, but the most notable ensemble was a large collection draped off his head and woven into his hair. The patterns of alternating white-and-black rectangular ceramic beads emphasized his ebony skin and honest face.

Atop his head were shrunken animal stomachs and other such vessel organs, extracted, I assumed, according to ritual proceedings. I thought I saw a small mirror and bone tucked away, also. Leather thongs wrapped around wherever possible. I estimated they had been in position for some time. Around his neck hung his cow-horn identification.

I asked him about "throwing the divining bones." I was thinking about my dissertation.

"I do not know answers to questions about life," he replied, "or about what I'm doing and why I live and die. No one else I have met does, either, although some pretend to know. I look for answers to questions that affect people's lives here. That makes sense to me—answers let me and my people live. I talk to my ancestors and throw the bones. I get answers I need. Who else can give the answers needed? No one."

I knew Africans practiced ancestor worship. Some, like the Rhodesian Shona, believed all illnesses had a cause that could be found only in the spirit world.

"And the bones you throw? Are they really just bones?"

"We call them divining bones," he said, chasing a fly away with his cow's tail fly swatter. "Each sangoma collects his own, and they may be anything he thinks is important or represents a significant time or accomplishment in his life. Or a patient of his might give him something as a reward for helping him. It could be a bullet or domino, shells, nuts, dice, or a part of an animal. No matter what it is, it can become part of the sangoma's divining bones."

"How do you interpret the thrown bones? What do you say to someone who is sick and wants to be cured?"

"I instruct my patients based on how the items land when I throw them like dice, as well as what my ancestors tell me. For example, I may tell a patient to sleep for one day or walk to a certain point."

"How do you know when you are speaking to an ancestor?" I asked.

"Sometimes my patients speak to me in a language they have never spoken before. Then I'm sure I've reached an ancestor. He is speaking to me from somewhere."

I looked at his garb and realized that it was an integral part of his ritual. It added to his authority. I thought about the allopathic physician's white coat and stethoscope. A patient sees those and believes help is on the way. A sangoma proceeds on the basis that if the mind is healed, the body will be okay. The Western doctor is more apt to treat the ailment and assume if that's remedied, the mind will be okay.

The healer showed me a three-foot-tall wooden statue he had carved some years before. It was of a thin female leaning somewhat forward with a straight back and jutting breasts. Wrapped around her waist was a short rawhide dress. The leather was stiff because the curing procedure had been crude.

Her base was uneven, as are most that stand on sand when they are carved. If placed on a flat surface, it would fall. The sangoma had no other statues that I could see and was not selling such things. It seemed it was his creation for himself; he wanted to carve it like some painters want to paint. I liked him; he was genuine and I wanted to bring something of his back to Durban with me. Not knowing much about carvings, I asked him if he wanted to sell it, thinking it would help him and me.

He paused. "For money?"

"Yes," I replied, "I don't have anything to trade."

"Okay," he said and asked a price of ten dollars which, given the circumstances, seemed fair.

"You must take it quickly," he said. "Or my wives will become too upset. *They* think it brings them luck."

I again thought about my dissertation and decided to take the statue for luck.

Indeed, when the women saw me carrying it to the Land Rover, they began to wail in protest. I almost left it there, not wanting to cause disruption. I looked back, and he pointed for me to go.

Some ten years later while I was in the United States, a dealer in African art offered me $10,000 for it. But to this day, through the trials, tribulations, and travels of my life, I've kept that statue. I've lost track of countless other things, but never that. Each time I admire it, I remember the Zulu nation and Africa. I think about luck, and the sangoma who carved it, and always about my dissertation. Then I acknowledge my full, healthy life.

Some weeks later, the two other responses came in at almost the same time, one from Britain and the other from Australia. Each carried an apology, saying the professors had been on vacation.

I had bitten my fingernails short by then. It seemed a long time between the delivery of my first review and that of the second and third.

Both had comments, and both provided the magic sentence:

They had no objection to awarding a doctorate degree to me. One letter said, "Your conclusions ring true, like the sound of the hard metabasite rock you hit." I've never forgotten that.

The bell of victory had pealed.

* * *

My first opportunity to notify the geological community of my findings came when I presented a paper at the Eighteenth Congress of the Geology Society of South Africa. The congress was held in Johannesburg. Geologists from all over South Africa attended, as did interested scientists from all over the globe.

As the day of my presentation neared, the memory of the disastrous lecture I had given at UCT two years before haunted me. The image of Afrikaner Joost and Professor NASA was clear in my mind. The very thought made me feel clammy. It was like returning to the rack. I didn't want to go through that humiliation again.

I bolstered myself, knowing I had worked hard and others supported my conclusions. Given that, I was ready to go back into hand-to-hand combat.

Martin asked me to prepare the lecture and give him a trial run. He liked it and made a few clever suggestions for improvement, but it was to be my responsibility all the way.

The conference day came, and I sat in the auditorium with Martin and about sixty other scientists in the audience. A current member of the PRU preceded me in the program—none other than Alan, the student I'd met in the field just before I left Cape Town, and whom Afrikaner Joost had brought to the university. I listened while thinking of my own speech. He referenced Joost's incorrect findings as if they were true, as he'd suggested he would years earlier.

At last, my name was announced. I stood.

"Good luck," Martin mumbled, probably as nervous as I was. He had taken a chance accepting me as a student in view of the problems I'd had, and this was where the rubber met the road. I walked onto the stage and to the podium. To the side, a projection screen would display images of my slides.

I swept my eyes across the gathering. My gaze settled on Afrikaner Joost and Professor NASA, sitting close and whispering. *Jesus, they are smirking at me again.*

I began to speak. I introduced my field areas as background and noted the data I had collected. Soon I relaxed and my words came easily, flowing out in a logical fashion. I had practiced the speech many times, and I knew I was right. My voice was strong; and at six foot three, I dominated the stage.

The audience followed my reasoning that Joost's published conclusions had to be wildly at odds with reality. All was quiet; people gave me their full attention as I continued. Many of them had heard about the conflict. You could hear a pin drop. I gestured to underscore my points. Several times I stepped to the side of the podium to emphasize the slides projected on the screen. Still the audience sat in what seemed like rapt attention.

After ten minutes of speaking, I found it in me to look directly at Joost and Professor NASA. *If I could face a leopard,* I thought, *I can face them.*

I held eye contact until they looked away.

I concluded my presentation and asked if there were questions. There was not one. No objections, no request for clarification, nothing. Silence. I don't think anyone moved.

"Thank you," I said. It was like pulling a plug. The audience applauded without hesitation.

I walked back to my seat.

Martin sat straight in his chair and smiled broadly. "You're glowing," he told me.

After the program, the head of the geology department at UN-P,

Professor Don Hunter, came to me and extended his hand. "Nothing wrong with that speech," he said.

I'll never forget those moments. I had won.

As for Joost, he ignored me. He seemed almost indifferent. I got the strangest feeling he hadn't understood what I'd said.

"You know why he did that?" Joost apparently said to another geology staff member of his department when he returned to UCT. "He did it to get back at me."

Joost was still trying to avoid reality. He couldn't face that he was wrong. The final crushing weight of that immoral situation was finally lifted from my shoulders. My life could go on. The slate was cleared of what should never have been.

I submitted an article on my findings to the international journal *Precambrian Research*, whose external editors reviewed it and then published it in issue 13, 1980, pages 253–274. I fully named and acknowledged my friends as I promised; it would remain available as a reference forever.

That summer, I received my PhD degree at a ceremony at the University of Natal. The dean of the faculty of science tapped me on the head with his mortarboard, and with that tap, I held a degree that no one could take away from me—a dream begun in my youth, fulfilled.

* * *

During my years in southern Africa, I wandered in the Transkei, Cape Province, prohibited diamond-bearing areas, the Elephant Coast, and the Wild Coast. I mixed with the Xhosa, Zulus, witchdoctors, white traders, and Afrikaners. I filled a treasure chest with adventures, some of which I've shared. Many more remain locked inside me.

A few weeks after receiving my degree, I heard about an old abandoned mine hidden in the Transkei hills that might contain gold. I'd been accepted as a lecturer at a nearby university and had just finished teaching for the semester.

Ever dreaming, I packed the geology department Land Rover and headed off to talk to trader Henry about it.

Epilogue

Dr. Martin P. A. Jackson later joined the University of Texas at Austin, and over the course of more than thirty years became the world's foremost authority on salt tectonics. He founded the Applied Geodynamics Laboratory at UT Austin, which is widely considered the world's premier institution for salt tectonic research. Described as a brilliant man not afraid to take a chance, he was not only my advisor but also my friend. Much to my great sadness, he died May 31, 2016.

Patrick went on to become a well-respected senior geologist working for major mining companies. He chased gold, copper, lead, zinc, tin, and tantalite deposits over most of southern Africa until he decided in the late 1990s to immigrate with his family to Australia, where he lives now. He didn't complete all the studies he wanted to at the PRU, which bothered him. But someone told him, "Don't look back, you're not going that way," and that—along with his fine character—kept him focused on the present. We remain the best of friends.

Catina completed her MSc in geochemistry. She remains a well-known champion of the environment and women's rights in southern Africa.

George Swanson died some twenty-five years after I met him at the age of seventy-two, when he was still spry and agile. His son is

CEO of George Swanson Enterprises (Pty) Ltd, which continues to be located at the Swansons' Springbok home I'd come to know and respect.

Afrikaner Joost continued on a few more years until he retired. The PRU was then merged into the geology department at UCT.

Professor NASA returned to the United States, where he lectured in geology at a major university and, lugging around his list of publications, became an emeritus professor.

South Africa, after a long struggle waged by the African National Congress and other activists, saw discriminatory laws begin to disappear from 1990 onward until 1994, when all ethnic groups were allowed political representation in the country's democracy. Poverty and inequality continue to be widespread, but the country has the second-highest economy in Africa and is considered a middle power in international affairs.

Rhodesia's Robert Mugabe became prime minister in 1980 and president in 1987, a position he held until 2017. He was considered by many to be a dictator; his country has suffered profound economic and social decline.

Mozambique held its first multiparty elections in 1994. It remains one of the poorest and most underdeveloped countries in the world.

After graduation, I lectured for the next six years on subjects like metamorphic petrology and structural geology at the University of Durban–Westville in South Africa. Martin Jackson and I published a chapter focused on my work in the book *Precambrian Tectonics Illustrated* (E. Schweizerbart, Stuttgart, Germany, 1984), coedited by Alfred Kröner, University of Mainz, Germany. Martin, Don Hunter, and four others published the book *Crustal Evolution of Southern Africa: 3.8 Billion Years of Earth History* (Springer-Verlag, 1982), in which my findings were significantly noted.

After leaving South Africa, I returned to the United States and entered the business world. This led to prolonged environmental work in Russia and Kazakhstan in the wild 1990s when the Soviet Union disintegrated. I started my own company there in the early 2000s and ran it until I moved to Thailand in 2015 to continue writing this book and others concerning my adventures. After traveling for more than fifty years, in mid-2017 I returned to the United States with my Kazakh wife and child.

Reading Group Guide

1. The author tells us that his heroes were "men like James Fenimore Cooper, Jules Verne, David Livingstone, Mark Twain, and Sven Hedin." How do these men fit with the lifestyle George has chosen? Who are your heroes and how do they fit into your life choices?

2. The author vividly describes the surrounding landscapes throughout the book. Based on his descriptions, how do you think he felt about the African environment?

3. This story takes place in South Africa during the apartheid period, and the author has several experiences with the local rules and culture surrounding this. How did he deal with it? Do you agree with his reactions? Would you have dealt with this differently?

4. Although most of George's outside research was conducted in solitude, it is obvious from the text that he is curious about people and has a generous heart. What are some examples of where the text shows this side of him?

5. Many of his stories and descriptions have a humorous tone and may have caused you to smile or even laugh out loud. Which ones were your favorites?

6. Instances of George's curiosity and imagination are woven into this narrative from beginning to end. One example occurs when, while reading *The Lord of the Rings*, George discovers some termite nests, and he compares them to Hobbit homes. What are some other examples in the story where George demonstrates his vivid imagination?

7. *A Rock and a Hard Place* is unusual in that the author is studying for his PhD while experiencing these adventures. He shares many details about geology with the reader. What did you learn about geology that you didn't know before?

8. In chapter 6, George first discovers that the area he was supposed to be studying had no "zones," which refuted the previous researcher's claims. As he and Catina discuss this, she advises him to use caution, "They will not want you to show they were irresponsible to accept such erroneous work. Rocks can drop down well with no splash." Have you ever been in a situation where you had to avoid stepping on toes? How did you handle it?

9. In chapter 7, George hears a story about a Bushman being hunted and killed by an Afrikaner's father. What parallels do you see between this situation and American history?

10. In chapter 11, Patrick and George encounter a dying gemsbok, whom they name The Lady. Several themes run through this story, including compassion and the circle of life. What did you think about this story? Would you have put the gemsbok out of her misery? Have you ever had an encounter with a dying animal?

11. Throughout the book, George continues his research while deliberating how to confront Afrikaner Joost with his findings. What personality trait does he demonstrate with this story? How would you have handled the situation if you had been George?

12. *Bonus Question:* The author has left a very pointed message for one of his aggressors hidden as a character in his story. Only those intimately involved in his predicament would instantly recognize this clue. Who—or what—do you think this is? Can you identify it?

Acknowledgments

My highest respect, regard, and thanks to Professor Th. G. Sahama, Dr. M. P. A. Jackson, Professor D. R. Hunter, James McDaid, and Paula Cardoso. Educators and friends, what would the world be like without good people like you?

Years ago, Laurie Rosin reviewed a large, unorganized collection of my stories, suggested a direction to take and edited a first version of this book that she adamantly stated "should be in print." I thank Laurie for her unwavering encouragement over the years to include her exceptional editing ability. As I worked in various places around the world, my manuscript was edited a second time by Jennifer Barclay and, recently, putting beautiful frosting on the cake, by Elizabeth Brown (who has traveled in Africa herself).

Some of the other people I would like to thank who, each in their own way, have influenced my life and this work: Professors Alfred Kröner, Kalervo Rankama, Peter Fleming, John Gurney and Dr. Kari Kojonen, Michael Carr, John Lynham, Mary Ann and Bob King, Myles Worsley, Gulmira Kaliyeva, and my daughter, Katherine Schade, who provided intriguing comments on the manuscript.

My utmost appreciation goes to Greenleaf Book Group (Danny Sandoval, Jen Glynn, Carrie Jones, Chantel Stull, and Chase Quarterman) as they accepted my narrative nonfiction book within three days of receiving it and, proclaiming it a wonderful book, didn't stop from that day forward encouraging me to continue my efforts.

Last but not least, I thank Zhanar Zelt, who, as director of CabinNotes LLC and my wife, applied ability, cunning, understanding, and the utmost patience to guide and encourage me as I wrote day after day, year after year. I'm a very lucky man to have her and our lovely daughter, Elizabeth.

About the Author

George left his Buffalo, New York, home in the 1960s when he was seventeen years old to hitchhike across the United States and Canada on such famous roads as Route 66 and the Alaskan Highway (sleeping in corn fields and hedges and among huge pine trees). In the 1970s and 1980s, he continued, building a shack on the shores of the Red Sea (which was burned down by vigilantes), traveling the coastal roads of Yugoslavia (where he ate with gypsies), and venturing farther north to the Arctic Circle to see the Laplander people before encountering his greatest challenge in Africa. He worked as an environmentalist and then developed his own company in the lawless 1990s in Russia and Kazakhstan when the Soviet Union disintegrated. He holds five degrees, including PhDs from universities in the United States, Finland, and South Africa and has published scientific articles. He recently moved to Texas with his wife and child after spending two years in Thailand, working on his books.